BUSES ALEXANDERS REMEMBERED

Vol. 3. Bodywork 1943 - 1960

Two generations of Alexander single deck bodywork seen at Crieff.

Allan T. Condie

***Below:** The official photograph below is sheer indulgence on the part of the Author as it conveniently brings together a lifelong connection between Stirling and Leicestershire. Seen on the Drip Road at Stirling after completion we see another connection underlined. DUT127 was the last of 50 bodies built for Leyland on PD1 chassis in 1946, and is seen about to set off for PDI (Post Delivery inspection) at Leyland before taking up duties in the fleet of Allen's of Mountsorrel.*

Contents

ISBN 1 85638 032 7 Hardback **ISBN 1 85638 033 5 Paperback**

Design, Layout and editing by Allan T. Condie
Typesetting and reproduction by Rivers Media Services
Printed in England.

Allan T. Condie Publications

40 Main Street, Carlton, NUNEATON CV13 ORG. Tel./Fax 01455-290389

Introduction

Our third volume on Alexanders looks not at the vehicles of the operator, but those vehicles bodied by Alexanders Coachbuilding department from 1943-8 and Walter Alexander & Co (Coachbuilders) Ltd. from 1949 up to 1960. As with previous volumes in the series the emphasis has been on illustrations showing the vehicles in their working environments. An example of each large batch is included and between these can be found the 'one offs'. Additional photographs have been included to show particular details. One or two vehicles have eluded our attempts to get photographs of them; fortunately this does not detract from the comprehensive collection we have been able to publish.

Bearing in mind that Alexanders fleet at the start of the period covered by this book was the second largest provincial operator in Great Britain with a fleet of over 1500 vehicles, it was natural that Alexander bodywork should be prevalent in that fleet. The period of growth from 1945 onward saw even Alexanders coachbuilding side struggle with the SMT group fleets buying many vehicles with other makes of body. After the Nationalisation of the operating concern and the independence of the Coachworks, Scottish operators still predominated, but increasingly the style and shape of the 'Stirling' product was to be seen all over the United Kingdom. The willingness to serve customers large and small brought orders from odd corners which were to be repeated as time passed. However, it was in Scotland that, in the period covered, there were few fleets which did not have an Alexander bodied vehicle, either new or second-hand.

The close relationship with Leyland built up in the pre-war and wartime period was to remain but manifested itself in other ways, expressly in the bodying of demonstrators and show vehicles. Perhaps the most famous of Alexander types, the TS8 Specials and the Y types will have to wait for further volumes in the series; it is intended eventually to cover bodywork from 1960-85 in a single volume and the whole Alexander story before 1945 which will cover both operating and coachbuilding activities.

March 1998. *Allan T. Condie*

Acknowledgements

Once again without so many friends to help this book would have been impossible to complete. Again I have been able to identify the sources of many of the photographs but I apologise if any are incorrectly attributed.

David Bailey, John Burnett, Gavin Booth, Alistair Douglas, Robert Grieves, David Harvey, Doug Jack, Bob Kell, Neil MacDonald, Roy Marshall, Harold Peers, Ray Simpson, Raymond Thom, Jim Thomson, Peter Walton, Brian Wright, and Geoff Wright have all helped in some way, and again I must thank so many who dug deep into their photographic collections to provide the selection brought to you in this volume.

***Below:* A look inside the Glasgow Road, Camelon, factory around 1960 with a Tiger Cub for Alexanders under construction. Note also the 'non PSV' work in the background with bread vans on Ford Thames chassis for Land o' Burns Bakery. Further research needs to be done on this area of Alexander activity which may well be covered as an appendix to a future volume.**

The 1945 bus body as fitted to overhauled Leyland Lion Chassis.

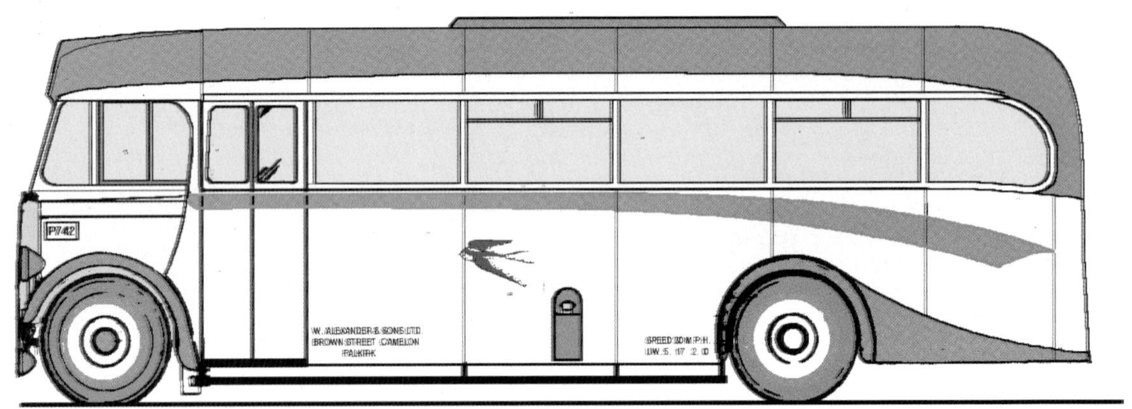

The first style of postwar single deck body fitted to Leyland Tiger PS1 Chassis

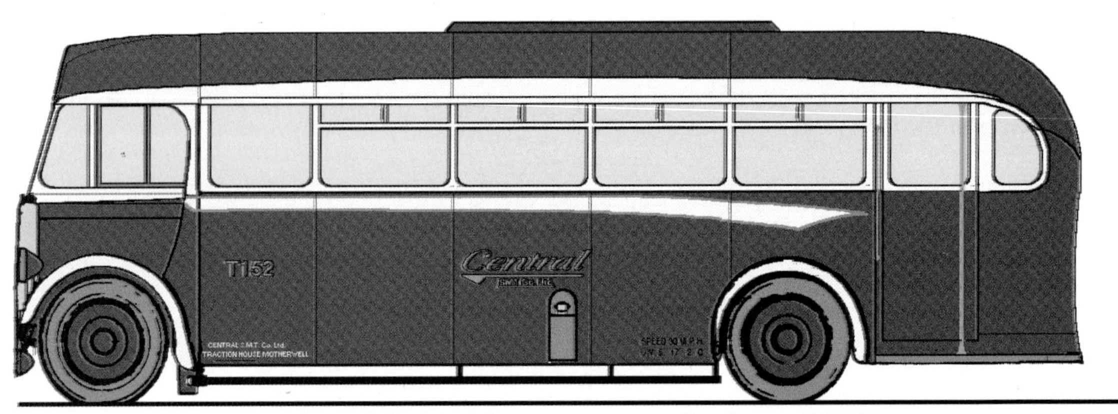

The second style of single deck body with rear entrance for Central SMT.

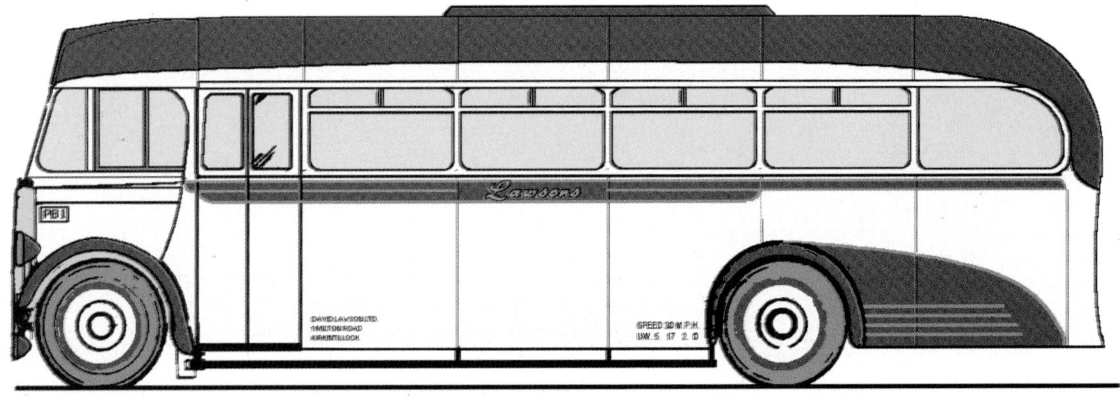

The final style of single deck body fitted to Leyland Tiger OPS2 chassis.

Leyland lowbridge body completed by Alexanders on TD7 chassis (SMT J66)

Wartime lowbridge body on rebuilt Leyland TS7 chassis for Central SMT

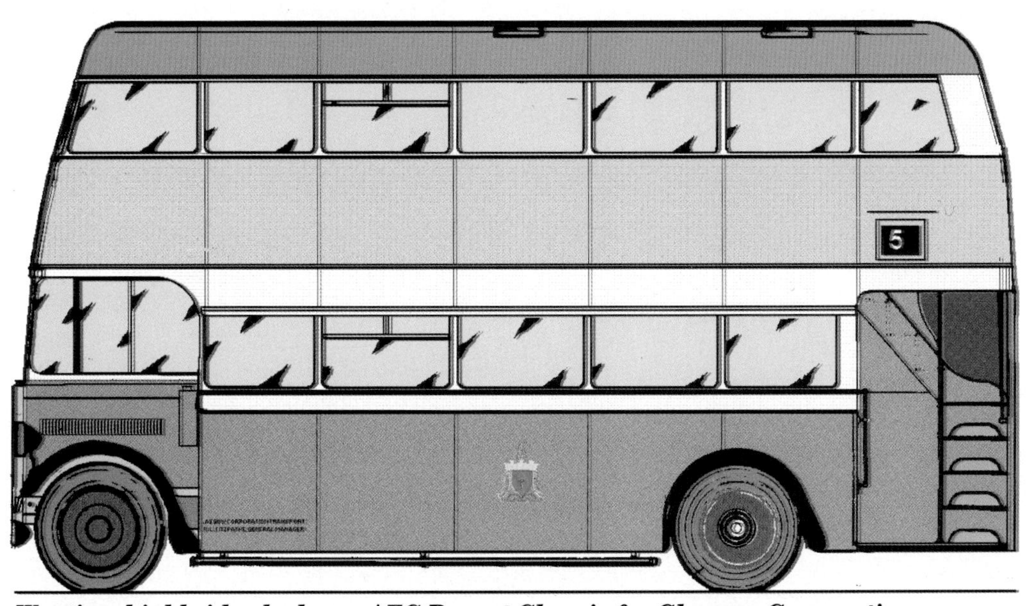

Wartime highbridge body on AEC Regent Chassis for Glasgow Corporation

Alexander Bodies from 1943

Alexanders bodybuilding department at Drip Road, Stirling had been engaged in building one design of body during the war years, a 53 seat metal framed double decker with strong affinity to Leyland designs of the prewar period, but with certain austerity features. These bodies were fitted in the main to rebuilt Leyland Tiger chassis from all the SMT group companies and some ex Regals from SMT themselves. There was also the odd rebodying job. Alexanders first double deck bodies had been built in 1942 on six SMT Tiger TS7 chassis rebuilt to TD4 specification and were virtually indistinguishable from the contemporary Leyland product. One more body for Western SMT was built to the same design and a number of TD7s arrived from Leyland in 'shell' form to be completed.

To provide extra vehicles for its own fleet in 1945 Alexanders cached a number of Leyland Lion chassis, mainly from Central SMT or Lanarkshire Traction, plus a few from its own fleet and having rebuilt the chassis with 6 cylinder engines of AEC or Leyland manufacture. Some of these were acquired at WD surplus sales in the form of twin units fitted to tanks, whilst others were culled from secondhand vehicles bought in for their engines. Although Alexanders had built using metal frames from the mid 1930s the 36 seat bodies fitted to these vehicles were even nearer to Leyland designs than hitherto, expressly in the use of Leyland style window pans. Whilst they were utilitarian in outline they bore considerable resemblance to the Alexander bus bodies of the mid 1930s.

Leyland's bodybuilding department found itself stretched at the cessation of hostilities and chassis production of its new PD1 double decker was exceeding the capabilities of the body department, which had been unable to restart production as the South Works was still being employed on war work. As Alexanders had acquired from Leyland during the war a number of jigs and tools, including a machine for forming the sections of metal framed bodies, they contracted to build bodies for Leyland. The first intention was for 50 lowbridge bodies which most likely would have been to Alexanders own wartime style. The order was changed to 50 bodies of highbridge specification to Leyland's own postwar design and were virtually indistinguishable from the Standard Leyland product.

The first job requested of Alexanders was a body for the prototype TD9 which later became CVA430, the prototype PD2. That was in January 1945 although the vehicle itself was not complete until 1946. The close relationship between Alexanders and Leyland provided for a body which to all intents and purposes had come from Leyland, the drawings for the postwar Hybridge design having been the first job executed by Leyland's design department.

Most of the remaining vehicles involved were supplied to Northern Municipalities. 50 bodies were delivered and it would appear that further work was transferred for assembly by Lancashire Aircraft, as Alexanders needed any spare capacity to cope with bodies for themselves.

For their own use, Alexanders continued with the wartime design of double decker body, slightly refined, and fitted it to both Leyland PD1 and AEC Regent chassis for the SMT group. It may have been a case of using up existing parts, but there was a shortage of skilled labour, expressly panel beaters, at the time, and the front and rear domes of the wartime design were easier to construct.. Another theory as to the perpetuation of this design may have been that Leyland had not issued drawings for the lowbridge version of its postwar body until April 1946. The first body, completed on Titan PD1 chassis in October 1947 caused a problem in that it failed its tilt test - the fitting of a stabiliser to the chassis rectified this.

The first postwar single deckers reverted to a design not unlike the prewar coach body as fitted to the TS8, but with detail differences. One idea which was not pursued was the 'special' concept; no doubt that Leylands had enough on their plates turning out standard PS1 chassis let alone do a 'special' version for the SMT group. Indeed the first postwar single decker of this type was built on an AEC Regal chassis and became Alexanders A28. Due to postwar shortages and the need to target materials for metal framed double deckers some of these bodies were timber framed. Examples of this body went to all the members of the group and Starks of Dunbar.

1948 saw the introduction of a new style of double deck body, which was almost identical to the Leyland lowbridge body of the time, There were subtle differences, as Alexanders tended to use flat horizontal panel strapping whilst Leyland used half round section and the PD1 variant had a full offside front mudguard. The emergency exit at the rear of the upper deck was the immediate give away, but early bodies even had the raised belt-rail mouldings on the lower deck. This body was further refined in 1949 when the window glazing was changed with flush alloy window pans becoming the order of the day. This move mirrored Leyland's development of their Farington style of body, the essential move to simpler glazing also being a weight saving exercise.

In the meantime, important changes had been taking place to the structure of the company. The 1945 Labour Government were pursuing a policy of state ownership of transport, and on nationalisation of the railways in 1948 the shareholdings of the LMS and LNER Railways in the SMT group, of which Alexanders was a constituent member, passed to the British Transport Commission. The SMT group sold the remainder of its interests to the BTC in 1949, but the coachbuilding activity remained independent and a new Company, Walter Alexander & Co. (Coachbuilders) Ltd., was set up in which the Alexander family had a large stake.

Until now, the main customer base for Alexander bodies had been the SMT group, and this group, now called the Scottish Group of Bus Companies, continued to be a major customer. From this time forward there was an increase in business from operators large and small throughout the United Kingdom and beyond.

Whilst supply of bodies in 1949 continued to the former SMT group companies in the form of single deckers on Leyland and AEC chassis to a slightly more modern design, and double deckers which still had a close affinity to Leyland designs, new customers were the order of the day, with Edinburgh Corporation coming for new single deck bodies on Daimler chassis dating from 1932, six lowbridge double deck bodies on Crossley DD42 chassis for Cardiff Corporation, and 25 similar bodies for Ribble on prewar Leyland Titan TD4 chassis.

Glasgow Corporation had received the only highbridge utility style bodies on prewar AEC Regent chassis in 1944, and came back in 1949 for more rebodying work, this time on Leyland TD4 chassis when 23 1935 examples were given new highbridge bodies. Ribble came back in 1950 for another 30 bodies, again on prewar chassis, and there were 15 more bodies for Glasgow Corporation on 1937 Regent chassis, these being of interest in having rubber mounted glazing.

A further order for Glasgow on new Daimler CVD6 chassis accounted for 40 bodies in 1949-50 and these were to a new, rounded, four bay design which also appeared in lowbridge form on Leyland Titan PD2 and AEC Regent III chassis for Alexanders, and both new and older Daimler Chassis for Western SMT. Strangely Scottish Omnibuses and Central SMT never received any of this style of body. The new design was developed not only to bring a more modern looking vehicle on the scene, but also as a cost and weight saving exercise, the general trend in double deck design encompassing four bay construction. Although Leyland produced four bay bodies for the London Transport RTW class, they never offered such a design to provincial customers, although the departure of Colin Bailey from Leyland to Duple produced a four bay version of their curvaceous double deck body; the Alexander version wasn't quite as 'curvy'.

Single deck activity in 1950 centred on Tiger PS1 chassis for Alexanders and rebodying of Leyland TS7 and TS8 special chassis for Western SMT which ran into 1951, The era of the front engined half-cab single decker was drawing to a close, but further examples went out in 1951, six on Regent III chassis to Alexanders and two on Albion CX39N chassis for Western SMT; these were of increased length due to the amended regulations in force from 1950 which allowed 30' single deckers on two axles.

The first underfloor engined chassis were bodied in 1951, these being 30 seat coaches with toilets on AEC Regal IV chassis for Western SMT and Scottish Omnibuses for use on the overnight Glasgow-London and Edinburgh-London services. The last half cab single deckers (well, not quite!) were 20 bodies for Alexanders on Tiger OPS2 chassis dating from 1948 which were acquired as part of a cancelled export order. Although these bodies were longer than 27'6" the extra space was used to give more legroom rather than provide more seats.

Early Postwar lowbridge body on Leyland PD1 chassis for Alexanders

Second style of postwar lowbridge body on Leyland PD1 chassis for Alexanders.

Third style of lowbridge body (revised glazing) on Albion chassis for Western SMT.

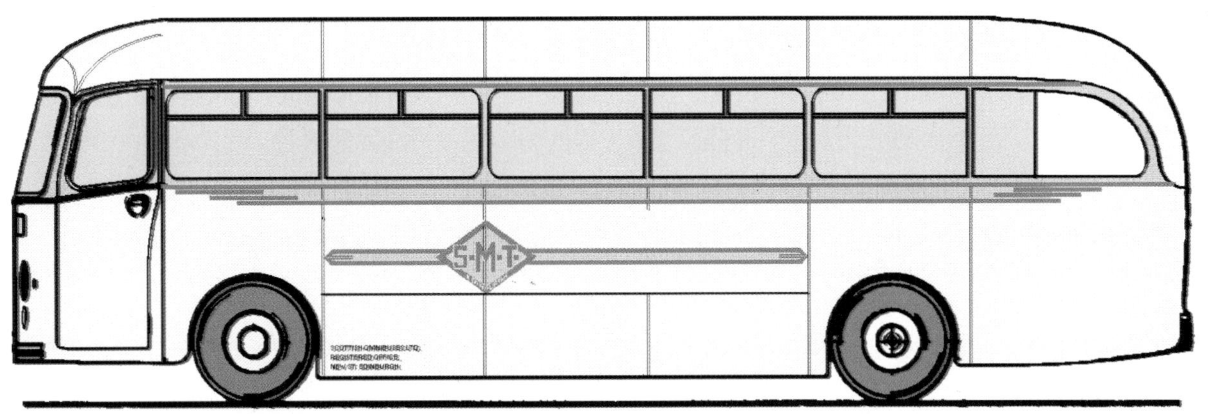

1951 London coach body with toilet for Scottish Omnibuses on AEC Regal Chassis.

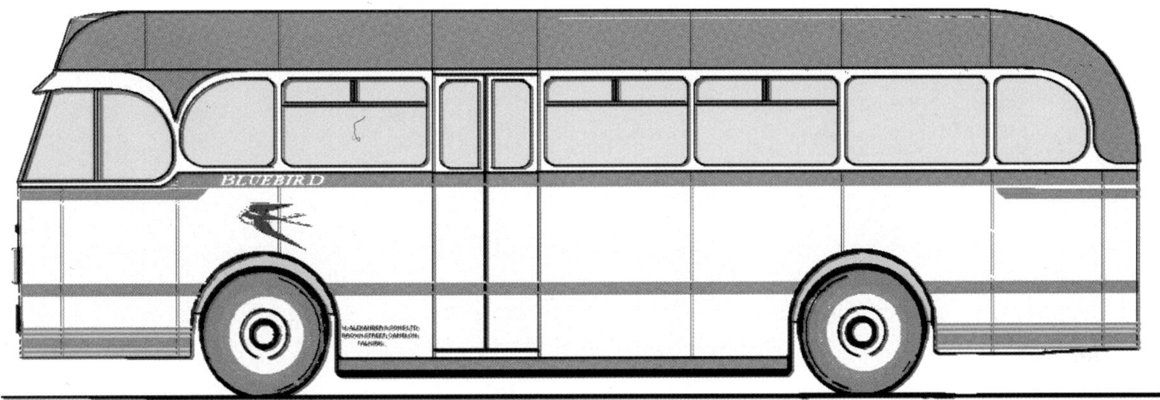

Coronation body on Leyland Royal Tiger Chassis for Alexanders

1954 Coach body on Leyland Tiger Cub chassis for Alexanders

1955 bus body on AEC Reliance chassis for Alexanders seen in Northern livery

The occasional 'one off' also made an appearance. For the 1949 Scottish Show Alexander bodied an Albion Victor FT39 with a unique style of full fronted body; Although displayed in Alexanders livery it was sold to Baird of Dunoon, and a second, similar vehicle went to Simpson of Aberdeen A Guy Otter appeared at the 1950 show and this was later sold to Monks of Leigh in Lancashire. Another interesting single order was for Leyland Motors on a Royal Tiger chassis and this was finished as a 31 seat rear entrance bus, entering service later with Edinburgh Corporation who sent it back to Alexanders in 1955 to be rebuilt as a 37 seat front entrance coach.

For 1952 the 'Coronation' style of coach body arrived and this was produced for only just over two years. Alexanders took this body on Royal Tiger Chassis from Leyland, whilst Western and Central SMT preferred Guy Arab UF chassis, an unusual variant being some bodies to the same style but with 'cut away' rear entrances in the Scottish style for Central SMT. A single order on Daimler chassis for Glasgow Corporation used the Coronation body shell to produce a 32 seat standee bus.

But the trend was towards lighter underfloor engined single deckers and to this end a 45 seat bus body which was to remain unique was fitted to a Leyland Tiger Cub chassis, and registered EWG240. It became a Leyland demonstrator and was later sold to Starks of Dunbar. In the event two distinctive designs for underfloor engined chassis came in the mid 1950s, the 1953 coach with its curved waistline and half centre bay carrying the emergency door on the offside, and the 1955 service bus whose side window layout owed something to the body on EWG240 but the front end was much plainer.

The only trolleybuses ever bodied by Alexander came in 1953; these were 5 Sunbeam F4A chassis for Glasgow Corporation and featured the then current 'hybridge' style of four bay body with deeply domed roof which was to remain an Alexander characteristic until the mid 1960s. There were also 16 bodies of similar style supplied to Edinburgh Corporation in 1954 on wartime Daimler chassis, and West Bromwich Corporation came for similar rebodying work on Daimler chassis taking 7 in 1953 and one more in 1954. In this case five bay construction was employed.

Whilst the supply of standard designs was preferred, some operators came with special requirements. Midland 'Red' took 11 unique coach bodies on their own C4 chassis in 1954, bodying the prototype themselves, and Glasgow Corporation took Leyland and AEC double deckers with bodies to Weymann design in 1955 and 1956. Another new customer was Liverpool Corporation who came for 60 bodies on Leyland Titan PD2/20 chassis in 1954 and 1955; these had some Alexander characteristics but the front and rear ends were unique, the final result being quite similar to other bodies supplied by Duple to Liverpool. 1954 also saw Leyland Motors cease building bus bodies and gave opportunity for Alexander to expand into new markets.

In 1955 the double deck body was updated with new front dome treatment. This design remained basically unchanged until 1966, and in due course 30' versions both with front and rear entrances were built. The lowbridge version for 1956 featured a rather upright frontal profile but this was replaced with a more stylish front dome design the following year.

The 1955 service bus body was built on both Leyland Tiger Cub and AEC Reliance chassis. Alexanders also took Park Royal bodied Monocoaches, a semi-integral design, and completed the bodywork on them. Sufficient strengthening was applied to the frames to enable the chassis to be driven from Southall to Falkirk.

In 1957 the two single deck designs were consolidated into a common shell for both coach and bus use. This was a 'straightened out' version of the 1953 lightweight coach design with straight waistrail. This body appeared on a variety of chassis; AEC Reliance, Leyland Tiger Cubs and Leopards and Albion Aberdonians mainly for SBG constituents, but others went to independents Baxters, Starks and Carmichael of Glenboig, with a solitary example going to Edinburgh Corporation. There were variations; the Baxters vehicles having a curved waistline but straight roof line, and Western SMT continued to take the original design with curved waist and roof lines.

Alexander were a natural choice to body the Albion Nimbus and one of the prototypes received a 31 seat bus body. There were, in all 30 Nimbuses and these went not only to Alexanders and Highland Omnibuses, but a few went outside Scotland. The final Nimbuses were given stylish bodies with glass fibre front and rear ends; the style was first seen at the 1959 Scottish Show.

It was late in 1957 that the first forward entrance double deckers arrived. Edinburgh took the first which had a unique front end and

Glasgow Corporation became one of the largest users of this type of body on both Leyland and AEC chassis.

Two rather attractive coaches resulted in the rebodying of two Daimler CVD6 chassis for Aberdeen Corporation in 1958; these were truly the last half cabs bodied by Alexanders. The following year Scottish Omnibuses received 20 unique coach bodies on AEC Reliance chassis with the emergency exit removed to the rear offside and five equal main window bays. The 1961 Tiger Cubs for Alexanders kept to the earlier design with half centre bay but had unique front and rear end styling in fibreglass. These remained the only ones of their type as a fire at Glasgow Road destroyed the moulds. But Alexander were getting increasing orders from BET group companies and this influenced the next designs as will be seen in a later volume.

Before reaching the end of our journey we need to be reminded that not all Alexander designed bodies were actually completed by Alexanders. In prewar and wartime days (1936-42) Alexanders were responsible for the supply of components to enable Dublin United to complete 135 single deckers at its Spa Road Works to a design virtually identical to that of the 1934-5 Leyland Lion LT5a rear entrance single deckers built for its own use. The final chapter in this story involved the two Titan PD1s for the Lough Swilly ordered as part of the Leyland sub-contract in 1946. These were shipped CKD and finished off at Spa Road. As DUT and later CIE built bodies using Leyland designed components this would be no problem.

Glasgow Corporation had always aspired to be a bus bodybuilder, but legislation prevented the complete construction of bus bodies at the Coplawhill Car Works. In 1946 they applied to Parliament for powers to construct trolleybus and motorbus bodies and this was granted despite opposition from the SMMT. Bodies to Alexander designs were completed on 50 Leyland PD2/24 chassis in 1958 and a further 25 in 1960, whilst 25 PD3/2 chassis were similarly constructed using Alexander parts in 1961.

Our final design to be considered is the early Atlantean, first seen on Glasgow's LA1 at the 1958 Commercial Motor Show. Similar bodies went to Northern General Group Companies and Newcastle Corporation, with a solitary example to Belfast Corporation.

Aerial View of Works, Falkirk, 1959

1950 lowbridge body on Leyland Titan PD2 chassis for Alexanders

1953 highbridge body on AEC Regent III chassis for Dundee Corporation

1953 5 bay highbridge body on Daimler CWA6 chassis for West Bromwich. (rebody)

1957 coach body with straight waistline onLeyland Tiger Cub chassis.

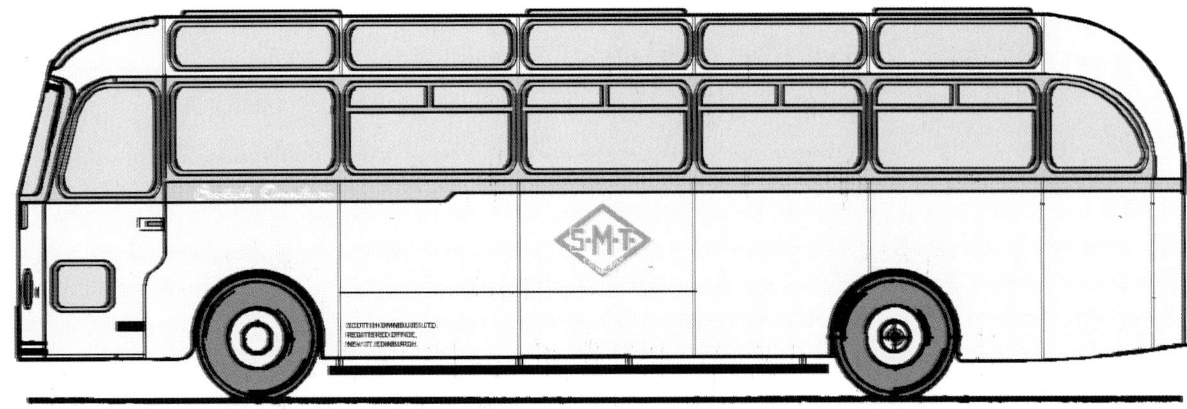

1959 coach body only supplied to Scottish Omnibuses on AEC Reliance chassis.

1959 coach body with curved waistrail supplied to Baxters of Airdrie.

New body on 1947 Daimler CVD6 chassis for Aberdeen Corporation in 1958

Studies in rear ends 1

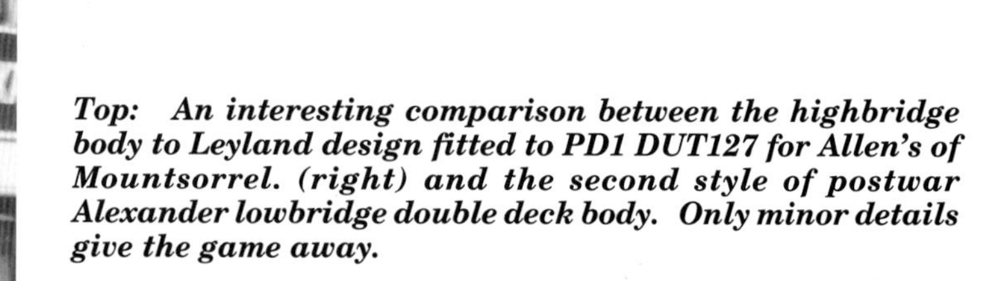

***Top:** An interesting comparison between the highbridge body to Leyland design fitted to PD1 DUT127 for Allen's of Mountsorrel. (right) and the second style of postwar Alexander lowbridge double deck body. Only minor details give the game away.*

***Left:** A rear view of the early postwar Alexander double deck design in this case fitted to a SMT Regent.*

***Above:** The 1950 design of rear end as exemplified by a West Bromwich Daimler.*

Studies in rear ends 2

Left. The rear view of the second type of Alexander front engined single deck body as fitted to Stark's Leyland Tiger PS1, L11.

Right: The unusual rear end on prototype body fitted to EWG 240.

Left: The rear of the 1955 bus design, in this case another Stark's bus, a Tiger Cub.

Right: The 1954 coach but with a difference! Barton 841 shows the rear destination equipment specified by this operator. The coach is on a visit to Scotland on a football hire..

1958 67 seat lowbridge body on Leyland Titan PD3 chassis for Alexanders.

Forward entrance highbridge body on Leyland Titan PD3 chassis for Edinburgh

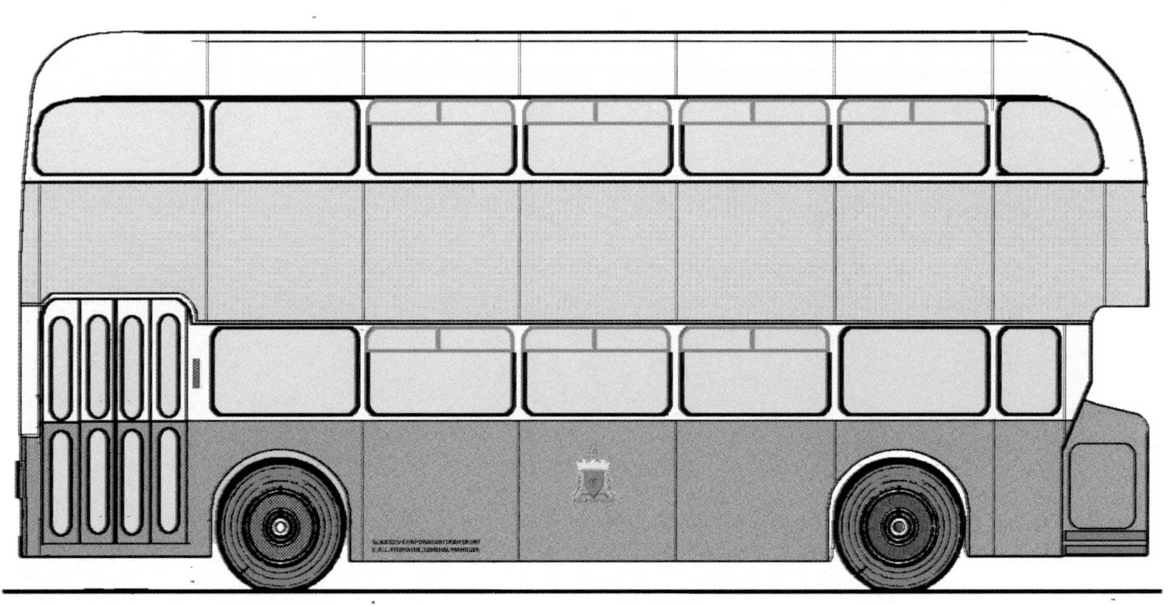

First Leyland Atlantean body for Glasgow Corporation Transport (LA1).

1. R332, seen here on the 'rural' part of the Falkirk circular, was one of 6 Leyland Titans delivered to Alexanders with bodies in shell form and completed at the Drip Road coachworks. Some of these vehicles had sliding ventilators from new.

2. The first complete bodies to be built by Alexanders from Leyland components were six 1942 examples fitted to the chassis of 1936 Leyland Tiger TS7s. They were body numbers 2564-9 and J61 is seen here in St. Andrews Square Edinburgh in June 1950. A feature of these vehicles from new was the lack of the characteristic lower deck belt rail pressings as used on true prewar Leyland bodies.

3. SMT J66 was an 'odd man out'. It was a lone example supplied to SMT in 1942; this body too was completed at Stirling. J66 survived in open top form as seen here outside Buchanan Street Station in Glasgow, and is now preserved with the upper deck restored - the components came from an ex Ribble PD2. The final 5 TD7 chassis supplied to the SMT group also included 2 for Alexanders, one for Greenock Motor Services, and 1 for Central SMT; the bodies for these were shipped separately and assembled on the chassis at Drip Road.

4. Western SMT CS4498, A Leyland Titan TD4 dating from 1936, was another oddity, its body was a replacement for the original which was destroyed by fire in 1941. Although this body was given an Alexander body number (2661) the outline was pure Leyland and this was the last lowbridge body so produced until the full Leyland shape reappeared in 1947.

5. The first Utility body was built on the chassis of P208 which had been rebuilt to TD4 standard at Brown Street before being sent over to Drip Road. Apart from the front and rear domes rear platform details and cab front the bodies were very similar to Leyland bodies in construction although austerity details were to be found inside. R365 was one of those bodies built on chassis which came from David Lawson Ltd., and is seen in Arbroath.

6. Nearer to home. Stirling's R465 was formerly P313 and was one of the last rebuilds supplied for 'home use'. It is seen here in Stirling not far away from its birthplace.

7. ***Glasgow Corporation sent 10 AEC Regent chassis dating from 1930 to Alexanders in 1944 and these received the only highbridge bodies of the wartime style. 270, which was new in June 1930 with a Cowieson body is seen in rebuilt form. Note the narrow bay behind the front bulkhead due to the extra length of the early Regent bonnet.***

8. ***SMT had both AEC and Leyland chassis rebuilt and fitted with double deck bodies. One of the AECs, BB17 is seen in St. Andrews Square Edinburgh, still in blue livery. There were 18 Regents body numbers 2869-86.***

9. ***The remaining 12 were Leylands, and J71 is seen here in Glasgow. The valance above the destination apertures hid slots for ventilation purposes; in time most vehicles lost this accessory.***

10. Western acquired two Leyland Titan chassis in 1943. These actually came from a batch of 6 TD3c buses bought by Alexanders from Sheffield in 1941, none of which were operated. They were sold to Millburn Motors in 1942 and the two which passed to Western SMT were sent to Alexanders to be fitted with body numbers 2801 and 2811, these being in the middle of the main batch for Alexanders own fleet in 1943.

11. Western didn't get the rest of its batch until 1944-5 when 50 Tigers were rebuilt at Kilmarnock to take new double deck bodies. This is CS5340 which received body 2942 in December 1945. It is seen in Ayr on a local service.

12. Central SMT received 30 rebuilds, and the first double deck body was supplied on L180 seen here at Traction House. It had been Tiger T77, and the body number was 2960.

13. Alexanders culled a total of 25 Lion chassis, from their own fleet, and from Central SMT and Lanarkshire Traction, and fitted them with this design of 35 seat body, the construction of which was very much based on the Leyland double deck design, as can be seen by the window pans and under the skin the use of diagonal cross bracing of the frames. P706 had started life as N218 in the Alexanders fleet in 1935 with a 36 seat front entrance body and received a Leyland 8.6 litre six cylinder oil engine in preparation for rebodying. The location is the top of Killermont Street, Glasgow, in front of Lyburns the Potato Merchants whose premises backed onto Buchanan Street Goods Station.

14. P725 was one of the 1945 rebuilds which operated on Perth City Services and it is seen here in King James Place about to turn up St. Leonard's Bank en route to Craigie and Darnhall. Originally red with "Perth City Transport" lettering by the time of this June 1950 photograph the normal Alexander fleetname is in place.

15. R257 was new in 1940 with a Leyland body but was one of three Titans whose original bodies were destroyed by fire during the war. It received body number 3006 in March 1946. The other casualties were R269 and R276 which received body numbers 3005 and 3007. R257 is seen at Callander Riggs Falkirk.

16. Alexanders built the body on the second prototype PD1 chassis for Leyland Motors. This was converted to PD2 specification and received the registration number CVA430 when with Central SMT. It was extensively tested by all 5 SMT group companies and also demonstrated to numerous potential customers in 1947. It is seen here in Halifax. The vehicle was dismantled in 1948 - it is just possible that the body was re-used but there has been no documentation found to prove this. This body was also unique in having a similar ventilation slot above the indicator as the wartime Alexander bodies. In fact it was originally to have a full front based on the Leyland Gnu design used by Alexanders but this idea was abandoned and a conventional half-cab fitted.

17. Just to prove that there was no difference between the Stirling built product and the home grown Leyland one, for comparison we illustrate Birmingham's first postwar Leyland, a PD2 built on chassis 470848, no doubt converted from PD1 specification. It entered service in October 1947 as No 296 with registration HOJ396 and despite first predictions it stayed with Birmingham until 1967. The seating layout was non-standard for Birmingham and it was the first body to feature the shorter cab front panel which exposed part of the offside front mudguard, a feature adopted on all Alexander bodies for their own fleet, even on PD1s, from the start.

18. In 1946 Leyland's bodybuilding department were unable to handle all the orders for new bodies so Alexanders were contracted to build 50 bodies on Leyland PD1 chassis. These were originally to have been lowbridge ones but the order was changed as explained below. Blackburn 72 was one of 6 supplied to that municipality and as with all the other bodies of this type the result was almost indistinguishable from the then current Leyland design.

19. Preston Corporation received seven out of a batch of ten from Alexanders. To the unknowing there was little to indicate that the bodies had come from Stirling. It is interesting to note that the three other vehicles delivered in 1946 had lowbridge bodies built at Leyland; despite Alexanders building this layout for their own use none of the 50 bodies for Leyland were of this type; at the time of the placing of the final contract with Alexanders the lowbridge body drawings were not available so the order was changed to highbridge..

20. Ornate shelters and gas lamps surround Darwen's No. 4, one of six bodied by Alexander. The prewar Titan in the background provides an interesting comparison.

21. Warrington's eleven PD1s joined an interesting municipal fleet which after taking these Leylands in 1946, later went to Fodens and then Bristols for chassis. In those days when Municipalities put all their contracts 'out to tender' it sometimes took the Transport Manager a deal of diplomacy and persuasion for the Transport Committee not to recommend the cheapest tender to the Council for approval.

22. Whilst most of the 1946 Alexander 'Leyland' bodies went to the North-west, three PD1s went to Tyneside. No. 30 is seen here in the attractive green and cream livery with the inevitable "Shop at Binns" advert applied to the front.

23. Back to Lancashire and its blackened red brick or stone, crown chimney pots and Victorian shop fronts let into terraced dwellings. Haslingden's bright blue and cream livery brightens up the scene as No. 25, the only one of its 1946 deliveries to come from Stirling, awaits its next turn of duty. The hills of Lancashire were not ideal for the staid performance of the PD1, once operators were given the chance to try a PD2!.

24. Midst the mills Accrington's unusual dark blue, red and black livery adorns No. 102 one of 7 with Alexander built bodies. Lancashire's mill towns were built not too far apart along the valleys of the rivers Ribble, Calder and their tributaries. There were a considerable number of joint services between towns and this added to the colour as each town had its own distinctive livery. It is interesting that most of the 50 bodies ran but a short distance from Leyland -somebody in the works may have been a little untrusting of Alexanders to do a good job so the vehicles would be handy for any rectification! History was to prove otherwise with the Stirling products having the same reputation for longevity as their Leyland brothers.

25. The furthest south that one had to go to find two of the 1946 bodies was Brentwood. The City Coach Co. received two; these would later be absorbed into the Westcliff on Sea fleet. LD1 is seen at Wood Green about to depart for Brentwood.

26. Our last visit to Lancashire for the moment finds us in Rawtenstall, about to take a trip to Burnley on No. 38, one of three with Alexander bodies. The rear end just visible is of a Leyland bodied TD4.

27. There had been Irish connections with Alexanders before, in the south, but two of the PD1s were partly assembled at Falkirk and then taken to Glasgow to be shipped to Belfast for forwarding to the Londonderry & Lough Swilly's works at Derry to be completed. In fact the LLSR sent the vehicles to CIE's Spa Road works in Dublin to be finished. These became Nos. 60 and 61 in that fleet, operating alongside some vehicles which received second-hand Alexander bodies from the SMT group single deckers recently rebuilt as double deckers, and some ex DUT double deckers with bodies built by the only other place to use Leyland components, Spa Road at Inchicore. Some former DUT staff were to find employment at both Leyland and Alexanders bodybuilding works in the immediate postwar era.

28. Only one PD1A was bodied by Alexander, the difference being in the Metalastik rubber bushed shackle pins of the chassis springs. It was also the only Leyland sub-contract order to go to a Scottish operator, McGills of Barrhead. It was body 3058. It is seen on home ground in Paisley with a Duple bodied Guy of the same operator behind.

29. The last of the contract, body 3059, went to Allen's of Mountsorrel, a Leicestershire independent, and received the registration number DUT127. This 'full frontal' was taken at Drip Road, Stirling and marked the completion of the 50 bodies for Leyland.

30 On home ground. in Loughborough DUT 127 rests before a return trip to Leicester. Allen's sold out to Trent in 1954 and the vehicle passed to Barton Transport.

31. After the 50 Leyland PD1s there was a gap of 16 in the body numbers which are believed to have been another batch for Leylands which were cancelled. Alexanders then turned to rebodying ten Leyland TD3 & TD4 chassis obtained as complete vehicles from Plymouth Corporation. R546 is seen at Alloa in company with an ex Chesterfield Corporation TD4 with M.C.C.W body.

32. The Scottish Group of bus Companies took what they could in the early postwar years and both Alexanders and SMT took AECs and Leylands. The first postwar Alexander single deck body was built on an AEC Regal chassis which became A28 in Alexanders own fleet, one of a batch of ten bodies from 3097-3106. Alexander's first 20 Regals had Burlingham bodies, as Drip Road were busy on the Leyland contract. A 30 is seen at Gairn Terrace Aberdeen when based at Stonehaven depot.

33. The next five bodies went to SMT and were identical, the design following very closely that on the immediate prewar Regals and TS8s but with revised trim and other details. B289 awaits departure from St. Andrew's Square, Edinburgh, although the driver seems to be inspecting the rear underside of the vehicle - a blown silencer perhaps!

34. Alexander's first Leyland PS1s were 12 new in March 1947 and when new carried fleet numbers P742-54. PA10. in original livery, is seen parked in Germiston Street, just outside Killermont Street bus station in Glasgow, still with original half-drop windows fitted to the early examples of the batch, no doubt to use up prewar stocks. The style of bus seat fitted to these 35 seaters was similar to that fitted to the immediate prewar bodies; note the way that the floor is sloped so that the seats over the rear wheel arch clear this.

35. Many of these single deckers had long careers, and NPA9 seen in Dundee passed to Northern in 1961, not being withdrawn until 1969 giving 22 years service. The half drops gave way to top sliders at the first body overhaul.

36. Just to prove that these vehicles were numbered in the P series when new PA25 as it later became is seen in this family snapshot when it was new and had P766 plates fitted. Note that depot plates had yet to be introduced. We don't know who the folks are - maybe someone reading this book may enlighten us.

37. SMT took prewar Regals in 1946-7, scrapped the original bodies, overhaule the chassis, and sent some of them to Drip Road for new Alexander bodies others receiving bodies by Burlingham. B164 was new in 1935 with a Cowieso 34 seat body. It is seen in St. Andrews Square bus station in Edinburgh abou to take a jaunt down the A1, with diversions, to Tranent. The SMT livery o the late '40s was two shades of green and cream. B164 lived at Dalkeith Garage

38. There were also some new Regals, 23 being delivered in 1947, and B297 is seen heading west along Princes Street. By now the dark green flashes have given way to cream. The A suffix after the fleet number shows that this was a New Street vehicle.

39. Alexander's first PD1s featured 30 bodies to what was basically the wartime design, but these shared lower deck window sizings with contemporary Leyland postwar bodies, with a wide pillar next to the front bulkhead and not to the rear as with the prewar Leyland body. Seen outside Larbert Road is RA11 which was new in January 1948, body 3182. Note how the Leyland cab structure sits uncomfortably below the top deck front. It is recorded that the first of these vehicles, RA7, failed its tilt test and stabilisers had to be fitted to the chassis to rectify this.

40. From RA31 onwards however, the design reverted to a very Leyland-like one and there were 25 of these, mainly on PD1A chassis but a few PD1s were sprinkled through the batch. The subtle differences from a Leyland body were in the use of flat panel strapping along the horizontal panel joins and the front offside mudguard.

RA42 is seen in the garage yard at Kilsyth.

41. Alexanders continued to supply the wartime outline of bodywork on AEC Regent chassis for SMT, later known as Scottish Omnibuses. Seen in Edinburgh in later years in the dark Tilling green adopted by SOL in 1965, is BB26, it ran until 1969.

42. By way of a contrast here is an earlier view of BB41B, in light green, heading east along Princes Street before finishing its run from Bathgate. It has been said that Alexanders were unable to use the later style of body as used on their own RA31-55 on AEC chassis for two reasons, one they were worried that Leyland might object, and secondly there was a shortage of skilled labour for panel beating, the wartime pattern domes were more easily fabricated.

43. More PS1 Tigers were delivered in 1947, FPA 35 being one of a batch of 15. Seen in Fife days at Kirkcaldy this vehicle has recently been overhauled in the body shop and fitted with rubber window mountings. The rear mudguard flash has also been eliminated. FPA35 was withdrawn in 1968.

44. SMT B61 was one of another 10 Regals rebodied by Alexander in 1947. It is seen in original livery of dark blue and cream about to set off for Newcastle on Tyne from St. Andrews Square in Edinburgh.

45. Northern's bus livery enhances the lines of NA40 parked in Seagate bus station, Dundee. Again this batch of chassis was split between Alexanders and Burlingham for bodying, A40 being delivered in August 1947 and running for nearly 21 years. The Regals tended to live in the Northern area of Alexanders.

46. 1947 saw a total of 49, Leyland PS1s emerge from Drip Road. Only 5 went to the Northern area where they spent all their lives. Here Forfar's PA40 reposes at Seagate Dundee, still in coach livery.

47. Starks of Dunbar were confirmed Leyland users, and because of their connections with SMT and the Edinburgh-Dunbar service their vehicles were painted in SMT livery. Stark's however had their own interpretation of this as can be seen on L10, A Leyland Tiger PS1 new in 1947 with body 3336, parked in St. Andrews Square bus station. Stark's Leylands tended to be tagged onto the end of Alexanders orders, indeed this vehicle broke into a batch of PS1s for that operator.

48. Central SMT received 4 Leyland Tiger PS1s in 1948. It has been a matter of speculation that these vehicles were actually diverted from Alexanders. T146 is seen at rest.

49. In 1948 Alexanders updated their single deck design and the result is seen here. PA67 was new in March of that year, and is seen in Dundee virtually as built and in original livery. The front end was tidied up and the 'peak' effect eliminated, the lower side windows were given radiused corners, and there were now sliding vents in all four side windows. The waistrail was also straightened and the rear end styling altered accordingly.

50. There were also 35 AEC Regals for SMT to the same design.

By the early 1950s half cab single deck coaches were becoming 'dated' and in order to prolong their usefulness various schemes were used. Scottish Omnibuses sent 15 of the batch of Regals to Marine Works who rebuilt them as seen here in this St. Andrews Square view in Edinburgh of B336. Others from the batch received new Burlingham 'Seagull' bodies. The supposed involvement of Dickson of Dunbar in this operation is questionable.

51. RA56-60 proved to be the last double deck bodies for Alexanders to be real Leyland 'look alikes'. In later years, now in Northern livery, NRA59, allocated to the Rosehearty, rests in Fraserburgh in the company of two later products of Walter Alexander (Coachbuilders) Ltd.

52. Central SMT L304-319 broke the mould as being Central's first Alexander bodied double deckers of the postwar period. Only the style of window glazing could distinguish this vehicle from its Leyland bodied cousins. It was a Titan PD1a new in 1948, and it is seen in Hamilton.

53. Alexander's RA63 is seen at Aberdeen. Whilst the Leyland outline was still obvious on this type of body, the revised window layout with flush window glazing was to remain a feature of Alexander double deck bodies for some time. The window pans were in Aluminum alloy.

54. PA76 is seen in Dunfermline and was new in July 1948. The only alteration made so far to the body is the reglazing of the destination blind apertures. This style of body omitted the flash behind the rear wheel arch.

55. The equivalent body supplied to Scottish Omnibuses on 36 assorted pre-war Regal chassis in 1949. They were originally seated for 30 but later received 36 bus seats from withdrawn vehicles. The mudguards are replacements, the chassis having been thoroughly overhauled at SMT's Marine works before being sent to Stirling for their new bodies.

56. Central SMT received 14 Leyland Tiger PS1s in 1949-50 and these were unique in having rear entrances in the traditional Scottish cutaway style. 34 seats were provided.

57. PA116 is seen in St. Andrews Square Edinburgh, having worked through from Alloa via Kincardine Bridge, the then lowest bridging point on the Forth. This 1953 view shows the trim applied for the Coronation; the side flash was painted red and a device consisting of Union Jack and Saltire surmounted by a crown was applied to each side of the body.

58. Greenock Motor Services was until 1949 a subsidiary of Western SMT, and in 1948 20 Titan TD3 and TD4 chassis were sent to Drip Road for new bodies, which were to the second style. These came from various sources including Sheffield and Oldham Corporations. One of the ex Oldhams, is seen as Western SMT 739 in the depot yard at Greenock. Note the 'Local' inscription above the destination used to identify vehicles on the ex GMS routes in the Greenock area.

59. Ribble sent 25 Titan TD4 chassis to Alexanders for rebodying in 1949. This first batch had bodies to 7'6" width. 2057 seen here at a Rally at Southport in the late 1970s was purchased for preservation by Harold Peers in January 1970. Harold still owns the vehicle, and advises that the bodywork is still in very good condition, a testimony to the workmanship of Alexander bodywork.

60. Glasgow Corporation sent 23 Leyland TD4c chassis to Stirling in 1949 for rebodying. The resulting vehicles still had a Leyland air about them but one interesting feature was that only 52 seats were provided. The chassis were new in 1935 with Cowieson bodywork. L23 is seen in Renfrew Street, the City terminus of the 47 route to High Possil.

61. Edinburgh Corporation had 7 Daimler COG5 chassis rebodied by Alexander in 1949 and these had the characteristic 'Scottish' style of cutaway rear entrance. A18 is seen at the foot of the Mound.

62. Scottish Omnibuses had 27 Leyland Tiger TS6 chassis rebodied at Stirling in 1949 and H77 shown here in Princes Street, Edinburgh, was new in 1933 with a Burlingham body.

63. Cardiff's choice of Crossley DD42 chassis provided a rare combination of Alexander Lowbridge body when they took delivery of 7 such vehicles in 1949. No. 42 is seen in the City Centre.

64. An unusual purchase by Western SMT were a batch of 15 Albion Venturer CX37S chassis and these entered service in 1949, having been turned out of Drip Road with the then current style of Alexander lowbridge body. AN597 is seen in Ayr.

65. A long way from home! Scottish Omnibuses last half cab bodies on new AEC chassis were supplied in 1949 and there were 20 of these Alexander bodied examples, a further 21 receiving Burlingham bodies. The location is Gloucester Green, Oxford and the antimacassars indicate that this is the 'two day' service from London to Edinburgh.

66. PS1s continued to enter the Alexander fleet and PA140 was delivered in September 1949. By the late 1950s many were repainted in bus livery, very often following refurbishment as has been carried out to body 3711, with the use of rubber window mountings.

67. Western SMT took 16 PS1s seated for 33 in late 1949/early 1950 and these introduced the penultimate style of Alexander body for forward engined chassis with radiused window corners top and bottom. Western didn't use route numbers so the front dome was of a different pattern. Note also the lack of opening roof and that Western specified bus seats!

68. Further PS1s for Alexanders in 1950 included a batch of 15 delivered in March and April. PA153 is seen when operating from the other garage at Dunfermline (Market Street) coded D2. It moved to the other Dunfermline garage on the closure of Market Street in 1961 and survived for another eight years.

69. Ribble's second batch of Titan rebodyings involved 30 TD4s and one of these is seen in Blackburn bus station. These bodies were to 8' width. These vehicles would have had longer lives had it not been for the service reductions of the late 1960s.

70. There were a further 3 Albions for Western SMT in 1949, and one completed the batch of 16 CX37s chassis. The other two were on CX19 chassis which had second-hand Burlingham bodies when new. 366 is seen after withdrawal at Nursery Avenue.

71. A further four PD1a chassis were bodied for Central SMT in 1950 and L317 is seen here in Glasgow.

72. Alexanders however by now had taken on the PD2 with its O.600 engine and MRB69 is seen in Midland days in Glasgow parked on ground around Buchanan Street which had been cleared ready for redevelopment. This style of Alexander body fitted to this batch of 26 vehicles was the last to retain Leyland influence.

73. As illustrated in "Alexanders Buses Remembered Part One" this Albion FT39N chassis had been bodied specially for the 1949 Scottish Motor Show when it was displayed with the number CMS920 in Bluebird livery. It was sold to Dunoon Motor Services in 1949 and re-registered SB8184. After a spell with Graham of Milton of Campsie it returned to the Cowal shore in 1955 to Baird of Dunoon with whom it is seen here. It went to Weir of Machrie on Arran in 1960, and passed through two more local operators, The Kyles of Bute Bus Company Ltd., and Chisholm of Kames, before being traded in with Don Everall of Wolverhampton in 1965 who scrapped the vehicle.

74. Glasgow Corporation took delivery of 40 Daimler chassis in 1951. 39 were CVD6 models and were fitted with Alexander bodies which were the first of the new four-bay style. The first, D21 is seen here at the Pollockshaws Road/Victoria Road junction at Eglinton Toll whilst employed on route 7 from St. Enoch Square to Kingsbridge Drive and is passing McNee's Lounge Bar.

75. The solitary Daimler CD650 delivered to Glasgow in 1950 received a similar body. In 1960 this body was transferred to the chassis of D66, which had originally been fitted with a 'one off' Mann Egerton body. It is seen in Clyde Street before turning right onto the Victoria Bridge.

76. Western SMT took 25 of the new style of bodies, in lowbridge form, on 7 Daimler CVA6 wartime chassis and 18 CVD6 chassis which had been ordered by Youngs of Paisley. 940 is seen in Paisley.

77. Glasgow rebodied 15 further prewar Regents in 1950 with Alexander supplying this style of 5 bay body of similar outline to the rebodied Leylands but with rubber mounted glazing. AR309 is about to return to Glasgow's Broomielaw from Pollok.

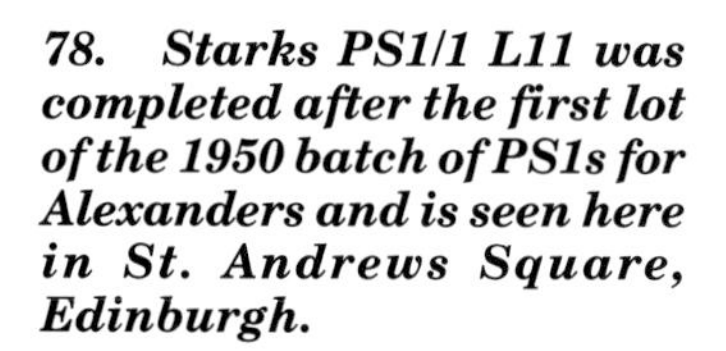

78. Starks PS1/1 L11 was completed after the first lot of the 1950 batch of PS1s for Alexanders and is seen here in St. Andrews Square, Edinburgh.

79. Western SMT had 8 TS7s which were new in 1935 and 1937 rebodied in 1950. One of the four 1935 examples new with Leyland bodies is seen here. Note the embellishment on the radiator; a feature of Western buses applied by individual drivers which mirrored the application of embellishments to pre-grouping locomotives by their drivers on the Glasgow & South Western and Caledonian Railways.

80. 7. The last 30 PS1s arrived later in 1950 and this brought the total operated by Alexanders to 196. The last chronologically had body number 3991 and PA183 was delivered in September. PA177, seen here in Stirling was new in September and was one of several with the same trim as the bodies for Western SMT being produced at the same time. This batch also featured radiused window corners top and bottom as first seen on the Western SMT examples and the return of the flash behind the rear wheel arch. Trim on the other examples was like Western SMT CS2027 in plate 79..

81. Another Albion FT39N was given an almost identical body to the 1949 Motor Show exhibit and delivered to Simpson of Aberdeen as DRS666. There were to have been 15 similar vehicles for Alexanders but the order was cancelled. The vehicle is seen operating for Smith of Grantown on Spey.

82. Western sent 20 TS8 specials for rebodying in 1950-51, leaving only one of this type with original body, the other 4 out of a total of 25 delivered in 1939-40 having been requisitioned for war service with the Ministry of Works and never returned. The result gave an indication of what postwar deliveries of PS1s might have been like if the policy of altering chassis to take high capacity bodies had continued after the war. Although Alexanders' own TS8 specials lasted into the 1960s with their original bodies, the reasons given for rebodying of these vehicles were the state of the floors and roofs; perhaps Nursery Avenue didn't have the capacity for heavy rebuilding work as was carried out on some of Alexanders examples to eliminate the opening roofs. AAG112 is seen in Dumfries.

83. A 'one off' body was fitted to this Guy Otter for the 1950 Commercial Motor Show at Earl's Court. Registered NTB403 it passed to Monks of Leigh in Lancashire but when seen here it was in the fleet of Hulley of Baslow in Derbyshire. It has since been preserved.

84. The 1950 style of Alexander lowbridge body looked very fine when fitted to the 20 AEC Regent III chassis supplied to Alexanders in 1951. These vehicles spent all of their lives in the Stirling and Falkirk area and RC6 is seen resting at Stirling bus station. With their preselective gearboxes they were ideal for the town services and 10 were allocated to Grangemouth, 6 to Bannockburn and 4 to Stirling.

85. Following on from the Regents were 6 AEC Regals which received the final style of Alexander body for half-cabs (almost!) and these vehicles were 8' wide and 29' long, but on 7'6" chassis. NA104 when new in 1951, as with all the others, was operated from Grangemouth on coaching duties, the manual gearboxes being somewhat of a let down after the preselectors of the AB and RC classes; they later migrated to the Northern area, with three at Dundee, two at Arbroath and one at Blairgowrie. NA104 was one of Arbroath's and ran until 1971.

86. *Western took similar bodies on just two Albion Valiant CX39N chassis in 1951 and BSD281, the first is seen in Ayr. Unlike the same style of bodies supplied to Alexanders, the Western examples seated 39.*

87. *There were 27 of these Leyland Titan PD2/3s supplied to Alexanders in 1951 and they were spread throughout the empire. Dunfermline's RB137 is seen after the 1961 split-up of Alexanders in Fife livery.*

88. *In 1951 Alexanders at last got round to bodying 20 OPS2 chassis which had been acquired from Leyland stock, no doubt at a bargain price. They were new in 1948 having been an order for New Zealand which was cancelled. Consideration had been given to their completion as double deck coaches!. They were however completed as single deckers but parts of them would see extended lives after 1960-61. With their 0.600 engines these were excellent performers and with 35 seats were widely used on extended tours before the advent of the Royal Tigers. Dunfermline's PB20 the last of the batch, is seen on home ground near Market Street depot having just undertaken a private hire.*

89. Edinburgh Corporation was experimenting with 'standee' type single deckers and JFS524, fleet number 801 was a Leyland Royal Tiger chassis fitted with a rear entrance 31 seat Alexander body. It was later rebuilt by the bodybuilder as a 37 seat coach. The vehicle was new as a Leyland demonstrator.

90. The first quantity order on underfloor engined chassis involved the bodying of 40 AEC Regal chassis for Scottish Omnibuses, the final 14 being diverted to Western SMT before delivery. There was no fixed design policy with underfloor engined bodies in the formative period from 1951-3, but the style of body seen here proved its worth on vehicles which regularly ran over 2000 miles a week between Edinburgh and London, for which duties they had 30 adjustable seats and a toilet; the hatch for emptying same can just be ascertained to the rear of the nearside.

91. There was nothing fancy about these vehicles, they were a simple, straightforward design which did the job well. One of those diverted to Western is seen after reseating to 40 in 1955 on normal stage carriage work at The Square, Cumnock.

92. There were no new double deckers for Alexanders in 1952, the ex London Guys being a possible reason. In 1953 the last of the PD2s arrived and these were PD2/12 models with longer wheelbase designed to take 27' bodywork. NRB 161 is seen in Northern days in Aberdeen. Fortunately this vehicle is preserved and now lives at the Scottish Bus Museum at Lathalmond.

93. Scottish Omnibuses received 10 of these AEC Regal IVs in 1952 with the same style of body as the London coaches but without toilets and fixed seats for 40. They ran virtually unaltered until withdrawal in 1966, being used latterly on stage and other bus work. The location, St. Andrew's Square in Edinburgh before the bus station was opened.

94. Alexander's first distinctive design for underfloor engined chassis appeared in 1952 and was christened the Coronation design just as the distinctive 1937 body on TS7 chassis had been prewar. The first numerically were 5 for Central SMT delivered in 1952 followed shortly after by another 5. They were on Guy Arab UF chassis. K35, the first, is seen here in later livery - note the use of rubber mounted glazing on the curved side windows. Some of this batch passed to Highland in 1964 but K35 was not one of these.

95. Western SMT also took the Coronation body on Guy chassis and some were 36 seaters with toilets for the London service, others like 961 were 41 seaters and it is seen here on stage carriage work in Kilmarnock working a local service.

96. Better known than any of the other 'Coronations' were those supplied to Alexanders on Leyland Royal Tiger chassis, the first of which PC1 is seen here at Aberdeen later in life in Northern's version of coach livery. New to Larbert Road in May 1952; note that these vehicles were delivered in time for the summer season of tours, PC1 went later to Aberdeen and finished off at Stonehaven being finally withdrawn in 1972.

97. One odd vehicle to appear in 1952 was this AEC Regal IV supplied to the Sutherland Transport and Trading Co. of Lairg. It is seen here in Inverness. The body style was that of the Scottish Omnibuses coaches of the same year.

98. Glasgow Corporation received this solitary Daimler D650H Freeline in 1953. The body shell was based on the Coronation coach design but there were seats for 32 and an open centre entrance, there being room for a large number of standee passengers. It is seen in St. Enoch Square.

99. The only trolleybuses ever bodied by Alexander were Glasgow's TG1-5 and TG4 is seen here in Cathedral Street which was a shortworking terminal point for route 101. The vehicles were built on Sunbeam F4A chassis and seated 62. They lasted until 1965-6.

100. FAG92 was unique. Built to the 'Coronation' design it featured several unique details, including the window treatment and a sliding entrance door. The special trim was applied for its appearance at the 1952 Earl's Court show. It became Highland Omnibuses K22 in July 1965 and is seen here heading north from Inverness.

101. The 1953 'Coronations' received different treatment to the bodyside trim and this wide band was originally painted red. PC55 spent its entire 18 year life in the Northern area and shows off the livery as applied to these vehicles by Northern. It is still being employed on long haul work and is seen in Edinburgh.

102. This unique bus body was built on Leyland Tiger Cub chassis and exhibited in Bluebird livery at the 1952 Show. It was sold to Starks of Dunbar in June 1953 , and is seen here at Dunbar.

103. Western SMT took the heavyweight bodies on Guy Arab UF chassis and Dundas Street bus station Glasgow finds 1021 in company with one of the later LUFs having just arrived from London.

104. There were an assortment of Wartime Guy Arabs in the Western fleet, including vehicles acquired from London Transport, Youngs of Paisley, and some delivered new. 28 were stripped of their original bodies, the chassis overhauled and sent to Drip Road for new Alexander lowbridge bodies in 1952-3. 1006 is seen in revitalised form in Dumfries. It was new as G369 with Weymann utility body, and ran for a time after acquisition in November 1951 with this body before being rebuilt.

105. The 1950s saw Alexanders gradually obtain more work from south of the border. West Bromwich Corporation sent 7 CWA6 and 1 CWD6 chassis for rebodying. The solitary CWD6, No. 132 is seen here in typical Black Country surroundings, the attractive two tone blue and cream livery brightening up the scene.

106. The last batch of heavyweights on Leyland Tiger Cub chassis numbered 31 for Alexanders and PC80, allocated to Stepps, is seen here in Glasgow in Midland days. After various livery variations these coaches finished their lives in the original style; an immaculate vehicle is seen here awaiting tour customers not far from Alexanders' Cathedral Street office. Two modifications were made to the vehicles during their working careers. The front visor has been removed and roof quarterlights fitted to the cove panels.

107. Scottish Omnibuses never operated any vehicles with the Coronation style body and their 1953 deliveries were unique. There was some relationship with the 1951 London coaches but the window pillars and glazing were unique to this style of body. There were 8 of these coaches for the London service with plain trim and B455 is seen in London. The London coaches clocked up around 6000 miles a week!

108. The other 17 were 38 seat touring coaches with enhanced side mouldings and roof quarterlights. Seen at Oxford's Gloucester Green is B465 operating on the more leisurely two-day service to London from Edinburgh via Carlisle.

109. Dundee Corporation took their first Alexander bodies in 1953 on 7 AEC Regent III chassis , and the last of the batch, 141 is seen here at Dundee's Crighton Street stance which was an overspill from those round the corner in Dock Street. They had the equivalent highbridge body to Alexander's own RC class and were mechanically similar. Dundee tended to use all manner of small coachbuilders as well as the better known ones in the early postwar period, but by 1953 some of those names such as Croft, Brockhouse, Cowieson and Scottish Aviation were about to become history if they hadn't already gone that way. Sister vehicle 137 has been beautifully restored and resides at the Scottish Bus Museum at Lathalmond.

110. Central SMT had a unique batch of buses with the Coronation body modified to rear entrance in the traditional 'cut away' style. There were 10 on Guy Arab UF chassis and K52 is seen operating a local service in Lanarkshire.

111. Liverpool Corporation Transport became an Alexander customer in 1954 with the delivery of 30 of these bodies on Leyland PD2/20 chassis. Built to Liverpool specification they used the basic four bay body shell but differed in front and rear end treatment. L79 is seen in later years in reversed livery operating the City Circle service launched in December 1965.

112. Dundee took one AEC Regal IV in 1954 with this unique 39 seat dual door body. It was used for flat fare experiments. The location is the Corporation bus stances in Dock Street with a more orthodox Weymann bodied Daimler behind.

113. A nearside view of Dundee 22 taken inside the depot shows the entrance and exit layout.

114 Edinburgh Corporation embarked on a rebodying programme of wartime vehicles and came to Alexanders for these distinctive bodies which were placed on wartime CWA6 chassis. They seated 58. 69 is seen here about to head along Princes Street towards the West end and Corstorphine, a route operated by trams until July 1954. The rebodying of these vehicles plus the purchase of ex London Guys which were rebodied by Duple was to enable the speeding-up of the tram replacement programme.

115. A new style of coach body was developed for the lightweight and medium-weight underfloor engined chassis. Alexanders provided four of these bodies on Leyland demonstrators in 1953-4. They were later sold off, LYS943 seen here was sold to Lowland Motorways in 1954. It was acquired by Scottish Omnibuses on the takeover of Lowland in 1958 and it is seen here in Stirling after becoming part of the Alexander (Midland) fleet in 1961.

116. FCS451 is seen here in Midland livery but its journey into that fleet also came via Lowland Motorways and Scottish Omnibuses. As SOL had no Tiger Cubs they exchanged the two ex Lowland examples for AECs AC21 and AC27. There were also a further two Demonstrators TTB80 and FMS245 with identical coachwork.

117. STJ989 was one of the first Alexander bodied Leyland Tiger Cubs and was new to Walls of Higher Ince, in Cheshire following a spell as a demonstrator. After a spell with Brunskill of Accrington it passed to A & C McLennan of Spittalfield in 1961 running for them until 1979. It is seen here in the yard at Spittalfield - the front grille treatment is not original!

118. Alexanders started to take Leyland Tiger Cubs with the new style of body in 1954, and MPD17 is seen later in life in Stirling, where it spent all of its 16 year existence. When new these vehicles were used mainly on private hire and tour work but they found their way onto stage services at times of vehicle shortages or holiday weekends. The roof quarterlights were a later addition. Other features of note with this batch of vehicles were the Eaton two-speed rear axles, air brakes, a new departure for Alexanders, and power was provided by Leyland 0.350 engines.

119. The Nottingham independent Barton Transport came to Alexanders for single deck bodies in 1954 and 676 was the first of a number of coaches of this style built in this case on Leyland Tiger Cub chassis. Note the curtains, antimaccassars, and the radio aerial.

120. Hutchison of Overtown, near Wishaw in Lanarkshire took two Leyland Tiger Cubs in 1954. The first is seen in Helensburgh. It was later sold to Rennies of Dunfermline.

121. Both KVA38 and sister KVA39 ended up with A & C McLennan of Spittalfield and the latter is seen here in Perth. The frontal treatment is not original.

122. Western SMT had a long association with the 1954 style of coach body and one of their first batch, 1094, based on Guy Arab LUF chassis is seen here at Nursery Avenue, Kilmarnock.

123. Liverpool's second batch of 30 Titans were delivered in late 1954/ early 1955 and L152 is seen here in traditional livery, heading for that Mecca of many Liverpool services, both in tramway days and after, the Pier Head.

124. An unusual order came in 1954 for 11 coach bodies on BMMO chassis classified C4. Midland Red's Carlyle works built the prototype body on the 12th chassis which in proportions was similar to the Willowbrook bodies on the C3 class. 32 seats were provided. These bodies had little in common with anything else that Alexanders produced at the time.

125. More Tiger Cubs were bodied for Alexanders and subsidiary David Lawson in 1954-5; out of a total of 6 delivered in mid 1954 two went to Lawsons, and one of these, PD33, is seen here, in Lawsons livery at Dundas St. bus station Glasgow. The balance of the order dragged on into 1955 and it was April of that year before the final vehicle was delivered.

126. Glasgow's D67 was a solitary example on Daimler CVG6 chassis It had appeared at the 1954 Earl's Court show and the body was unusual in appearing to have Weymann framing with an Alexander style front and rear to the upper deck. It is seen in Pollockshaws Road in the 1960s livery heading back into town on the 45 route from Rouken Glen which replaced the No. 25 tram in 1959.

127. 49 bodies of this style were supplied to Glasgow Corporation in 1955 and 1956 on AEC Regent chassis. The frames were supplied by Weymann. It is seen at the Renfrew Street stance on the 28 route from the City to Milton.

128. LVD263 was one of two Alexander bodied Tiger Cubs supplied to Hutchison of Overtown in 1955. It later passed to Gibson of Moffat in whose livery it is seen here.

129. Barton took another 3 Alexander bodied Tiger Cubs in 1954. Seen in Nottingham 733 is about to run deep into Midland 'Red' territory on the 3C to Swadlincote.

130. Edinburgh Corporation took three Tiger Cubs in 1955 for their City Tours work, and they were delivered in black and white livery. 819 is seen parked not far from St. Andrews Square in this shot.

131. Smith of Wigan took 5 similar vehicles, and one of these BJP 271 is seen in later life when operating for Viking Coaches of Derby and Burton on Trent. In the mid 1960s the author of this book used to feel 'quite at home' when this vehicle was in use on associated Company Victoria Motorways stage service from Measham to Burton on Trent, on which he travelled every Sunday evening.

132. SGK734 was an odd Guy Arab LUF supplied new to Kingston in London SW7. It is seen here when operating for Dance Motors. Note the opening windscreen which was a legal requirement on all PSVs until 1959.

133. Hall Brothers of South Shields took a solitary Alexander bodied Tiger Cub. It is seen here in Newcastle in company with various Northern group vehicles.

134. Western SMT took 30 Guy Arab LUF chassis and these were bodied in three groups. The first four were 41 seat coaches, the next sixteen were 44 seat buses and the final ten had toilets and 30 seats for the London service. All were outwardly similar in appearance, and 1104, one of the London coaches, is parked in Glasgow's Parliamentary Road outside George Boyd's, the Ironmongers. The normal departure point was Dundas Street bus station, but at busy summer periods the overspill loaded here; some prospective travellers appear to be heading for the bus and one behind whose reflection can just be seen in George Boyd's windows.

135. Alexanders took Guys as well, with 10 going the Fife area in 1955. GA9 is seen in its twilight years in Fife livery at Kirkcaldy Note the alterations to the front panel.

136. Glasgow Corporation became an increasingly more important customer for Alexanders in the mid 1950s, but this batch of 24 Leyland Titan PD2/25 chassis were bodied in 1956. They were built to Weymann design ostensibly on Weymann frames. L35 heads along Cathedral Street.

137. Barton 740 was an AEC Reliance and is seen here at Mount Street, Nottingham operating the express route 5 to Derby which was operated jointly with Trent Motor Traction.

138. Three Leyland PD1A chassis new in 1947 and fitted originally with second-hand bodies were rebodied in 1955 for Alexanders with what proved to be the last bodies of the 1950 style; although these were 7'6" wide. RA1 is seen here in Glasgow. It originally had a Roe body from a Ribble TD4.

139. The 1955 Alexander bus body appeared first on AEC Reliance chassis and the first 20 went to Alexanders, mixed in with deliveries of similar vehicles from Park Royal. MAC48 is seen after the Alexanders split-up, in Midland bus livery, when allocated to Balfron depot. The revised frontal layout was applied to improve engine cooling.

140. The final two Reliances from the 1955 batch went to Starks of Dunbar. A4 is seen in St. Andrews Square, Edinburgh before the opening of the bus station. Starks used their own version of SOL livery on vehicles used on the joint Edinburgh-Dunbar service for which SOL held the licences; note also the SMT diamond and SOL legal ownership.

141. The next batch to be bodied were Monocoaches for Alexanders, the running units and chassis frame being lashed up with temporary strengthening to get the chassis from Southall to Stirling, where the bodies were fitted. AC61 was one of 30, entering service in July 1955. It was based at Kilsyth, and is seen here in Dunfermline, in original livery, on the Glasgow -Dunfermline service.

142. 30 Tiger Cubs also received the 1955 bus body and delivery carried over to 1956. PD41, seen in the rain in Stirling sometime in 1962 has also had the front panel restyled and is in bus livery. PD41 actually entered service in January 1956 spending most of its life at Bannockburn until transferred to Cumbernauld in 1967 where it remained until withdrawal in 1971.

143. Baxters Bus Services of Airdrie, still independent in 1956, took two AEC Reliances with bus bodies, these being delivered in January and May 1956. No. 16 seen here was the first to arrive, and these buses looked rather smart in Baxters two tone blue and grey livery. It passed to Scottish Omnibuses in 1962, lasting until 1970. Only the registration number and different destination screens would tell it apart from the multitude of Reliances and Monocoaches operated by SOL.

144. Edinburgh Corporation set about replacing the last of their trams with this batch of 50 Guy Arab IV models with 63 seat bodies. 904 is seen operating the 19 circle in George Street with the Commercial Bank of Scotland in the background.

145. NSF543 was a solitary AEC Reliance new in 1955 and it also appeared at the 1955 Scottish Motor Show. It had 30 seats plus toilet for the London service but is seen here on more mundane duties in St. Andrews Square bus station.

146. Scottish Omnibuses received 25 Monocoaches in 1956; these were seated for 41 and were therefore classed as 'dual purpose'. The last of the batch B546 is seen in Edinburgh heading west for Ratho.

147. In 1956 the double deck body was updated, and the result is seen here. The upright frontal appearance was no improvement on the previous style. Central SMT received 25 of these bodies on Leyland PD2/20 chassis and L530 is seen here in Glasgow, crossing Jamaica Street, in front of Paisleys department store, heading east for the Lanarkshire hills and the mining village of Shotts.

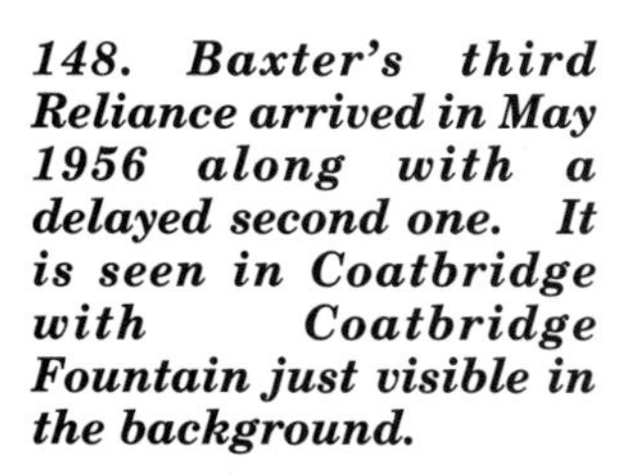

148. Baxter's third Reliance arrived in May 1956 along with a delayed second one. It is seen in Coatbridge with Coatbridge Fountain just visible in the background.

149. Western were next in line to receive the ugly 1956 body on Leyland PD2 chassis. 1225 is seen at Kilmarnock. Buses had a habit of finding parking spaces on waste ground, in this case caused by the town centre redevelopment.

150. After the PD2s came 13 Guy Arab IVs; these featured a less severe frontal treatment. 1257 is seen here in the 'Sou West'on the A76 heading for Dumfries. The telegraph poles and the lack of traffic are reminders of a past age.

151. Gash of Newark took a single Albion Nimbus in 1958 and it is seen here on home ground. Gash's were traditionally a Daimler customer.

152. Southampton Corporation took 3 Nimbuses, one of which was exhibited at the 1956 Commercial Motor Show. 256 is seen here on the One Man Operated service to Swaythling and S. Stoneham Cemetary.

153. Wiles of Port Seton took a solitary Nimbus and it is seen in its home village about to set out on a local run to Prestonpans.

154. There was one more Nimbus to this design which went to Leyland as a staff bus. It too finished up with Wiles of Port Seton as seen here.

155. The first 10 Tiger Cubs delivered to Alexanders in 1956 had 41 seats. MPD74 is seen in Stirling displaying Midland bus livery and the restyled front panel, along with the second style of Midland fleetname, before heading along the foot of the Ochils to Alloa.

156. The remaining 20 of the batch had 45 bus seats and PD82 is seen here in original condition and livery in Dunfermline. The style of seat used dated back to the 1930s and the TS8s.

157. Seen much further north in Inverness is Elgin's AC96, one of 19 Reliances delivered in mid 1956, all went to the Northern area and for some reason the body on the 20th chassis was not built. This vehicle spent all its life at Elgin, being later equipped for one man operation. It passed to Highland Omnibuses in 1975 but was not used. Our photograph shows the vehicle almost 'as delivered'.

158. Alexanders tended to add odd body orders onto the end of a batch and this single order for Greenshields of Salsburgh came at the end of the AEC Reliances for Alexanders. It is seen heading for Shotts on the Southern uplands between Edinburgh and Glasgow, midst harled council housing so typical of Central Scotland.

159. Another small order was for two 55 seat lowbridge bodies on Leyland PD2/20 chassis for Smith of Barrhead. Paisley Abbey provides a backdrop for the first, RGA632 as it prepares to head on the local service to Todholm.

160. Edinburgh Corporation's 1956 intake to eliminate the trams included 20 of these Guy Arab IVs with 63 seat bodies. 966 is seen heading up North St. Andrews Street before encountering the plethora of SOL vehicles in St. Andrews Square and heading along Princes Street to the West end.

161. The Nimbus body was supplied to Highland Omnibuses in the form of 6 29 seat coaches in 1956. The body had been restyled to incorporate a half bay in the centre, which with its straight waistline pre-empted the 1957 single-deck bus design. A3 is seen here.

162. Five similar vehicles arrived in early 1957, with all but one going to Alexanders associated Lawson fleet. The odd one out was N5 seen here on home ground at Stirling. These vehicles in part replaced the Bedford OBs on tour work; being the same seating capacity they could be charted for the same number of 'touristes'.

163. An unusual order for three Nimbus bodies came from Jacksons Motor Services of Castle Bromwich. Note the use of the dual purpose type of seat preferred by Alexanders in this vehicle but without the top rails.

164. *The final Nimbus went to Dunbarton County Council. It was not used as a PSV but employed on school and staff work. Note the jack-knife doors on this example; the only one of the later Nimbuses so equipped.*

165. *Ten More Guy Arab LUF chassis were bodied for Alexanders in 1957 and these went to the Fife area. GA12 is seen on private hire duties in Glasgow.*

166. *Central SMT took 12 lowbridge bodied PD2s in late 1956. Central territory penetrated the north bank of the Clyde with services inherited from the Clydebank Motor Co. and L559 is seen at Anderson Cross turning into the bus station.*

167. B456 was new in 1953 but was rebodied in 1957 with a very similar body to the 1955 show exhibit B547 following accident damage to the original in 1956. It too had a toilet and is seen here on a private hire near London's Wembley Stadium.

168. The last Monocoaches to be bodied by Alexander went to Highland and SOL in 1957. The first 6 went north and the first of Highland's half dozen, B1, is seen in Inverness. All 20 of the batch were dual purpose 41 seaters.

169. The remaining 14 went to Scottish Omnibuses and as befits its dual purpose status B574, based at Edinburgh's New Street Garage, is seen heading through Coatbridge with evidence of the trams still visible.

170. Bristol chassis and Alexander bodywork proved an unusual combination but one with which Western SMT seemed content. 1280 was one of 19 LS chassis bodied at Stirling in 1957. The location is Guildford Square, Rothesay on the Isle of Bute.

171. Glasgow Corporation took 50 Daimler CVG6 chassis in 1957 and these received Alexander 61 seat bodies of the style seen here on D136. The bus is just turning onto Jamaica Street Bridge with Paisleys, the famous Glasgow outfitters in the background.

172. Following on from the Daimlers came 50 Leyland Titan PD2/24 chassis with similar bodies. There were also a further 50 built to Alexander design by GCT. L68 is seen in Cathedral Street with Queen Street railway station below.

173. The 1957 lowbridge double-deck design had a better proportioned front to the upper deck and Western SMT took 15 on Leyland PD2/20 chassis. 1375 is seen on the 'long way round' service from Glasgow to Ayr via Troon.

174. The last vehicles built to the 1955 single deck design arrived in 1957, and these were 30 AEC Reliances with 41 seat dual purpose bodies for Alexanders. This batch were unique in not having conventional fairings over the wheel arches, but this set the trend for what was to come. These vehicles after the split were painted by Northern in coach livery, and Stonehaven's NAC107 is illustrated.

175. The 1957-8 bus body owed its origins to the 1954 coach design but had the waistline and roof line straightened. The first batch went to Alexanders on Tiger Cub chassis, again with 41 seats. PD125 is seen in Midland days heading north from Pitlochry up the A9 to Calvine Post Office, where the Struan run terminated.

176. More Leyland PD2/20s and PD2/30s for Central SMT came in 1957. L563 is seen here.

177. Members of the AA consortium from Ayrshire took three Alexander bodies in 1957. These were similar to the then Glasgow Corporation style but were fitted to three different makes of chassis. Tumilty received this Daimler CVG6 seen here in Ayr.

178. The Guy Arab II chassis used for Dodds DT8 came originally from CDR673, an ex Plymouth Corporation Roe bodied example new in 1943 and bought by Dodds of Troon in 1954. It was overhauled and fitted with a MkIII radiator and bonnet before being sent to Falkirk for rebodying. It is also seen leaving Ayr for the run up the coast to Irvine and Ardrossan.

179. The third body went on a new Leyland PD2/20 chassis. The Leyland was owned by Dodds of Troon. It too is seen on the Ardrossan service from Ayr,

180. An unusual order in 1957 were 7 bodies built on Albion Aberdonian chassis. These were to the 1954 pattern, and there appears to have been some shunting around with the deliveries at this time as Western took the first three as their 1384-6.

181. Alexanders received the other 3 which became NL1, NL2, and NL22. NL2 is seen here at Falkirk.

182. There was also one Albion demonstrator TGB752 seen here at Millburn Motors, the Leyland dealers. Albion came under Leyland control in July 1951, hence the fact that it is surrounded by Leylands.

183. Glasgow Corporation, with the tram replacement programme in full swing, took a further 50 Daimler CVG6 chassis in 1957 and sent them to Falkirk for bodying. With the Royal Infirmary in the background, D169 heads towards the City before crossing the Clyde and heading out to the south side.

184. There were also 50 more Leyland Titan PD2/24s . L153 is seen outside 'Woolies' in Union Street on the University-Mosspark Service which replaced the famous 'Yellow' tram route.

185. One solitary Daimler CVD6-30 arrived in 1958 and remained unique as the only 30' rear entrance double decker bodied by Alexander for Glasgow Corporation. It is seen at the Castlemilk Terminus of Route 34, and we are fortunate that this vehicle has also been preserved.

186. Edinburgh Corporation received this unique PD3/2 in 1957. It had been exhibited at the 1957 Scottish Show and was lent to both Central SMT and Western SMT in 1958. The unique glass fibre bonnet assembly was produced by Homalloy but was a 'one off'.

187. Alexanders had not taken any new Alexander bodied double deckers since 1953 and the 1958 deliveries were also the first to have 30' long bodies. This introduced a new style of lowbridge body seating 67 and there were 30 of these buses; they also introduced the BMMO style 'tin front' to the Alexanders fleet. Their extra capacity was welcome on suburban routes out of Glasgow and RB175 is seen in Glasgow's Dundas Street bus station.

188. A 'one off" Albion Aberdonian went to Edinburgh Corporation in 1957 as a 43 seat bus,. It is seen in original condition on trade plates before exhibition at the Scottish Motor Show that year. 822 only operated as a bus for a short period and it was repainted into the Edinburgh Corporation black and white livery reserved for coaches and had the front indicators removed from the roof. It then had its seats removed and was used to carry works of art to Italy for an exhibition in Nice.

189. On return the vehicle was rebuilt with 37 coach seats but spent some considerable time with its seats removed when in use as a mobile publicity vehicle on four separate occasions. In 1973 it again saw conversion into a mobile office for sporting events. It is seen here on one of its spells on City Tours work at the tours stance on the Mound.

190. Another odd chassis from Albion went to Scottish Omnibuses in 1958. This had been exhibited at the Scottish Motor Show in 1957 and it went eventually to Alexanders as their NL23, becoming NNL23 in 1962. It is seen in Aberdeen, its home base.

191. Western took more Bristol MWs in 1958 and with Alexander bodies which continued the 1954 style; indeed Kilmarnock standardised on this design when other Companies in the Scottish Group were taking the straight waistline version. 1402 was seated for 41 and is seen here on stage carriage work in Kilmarnock.

192. Western SMT's first 30' double deckers had Northern Counties (NCME) bodies; the first Alexander bodied examples arrived in 1958 as part of a batch of 43, the remainder being bodied by NCME. All the 30 footers featured platform doors and 1458 is seen at Ayr bus station.

193. Alexanders took 16 AEC Reliances with the new style bus bodies in 1958, seated for 41. This gave them dual purpose status. Peterhead's AC134 is seen in Northern days at Aberdeen.

194. Barton Transport took 5 more AEC Reliances in 1958, specifically for touring work. 773 is seen here fully 'decorated' for an extended tour to Switzerland - or is it just a pose for the camera?. The perspex panels for a tool kit were unique to Barton; note also the side luggage locker.

195. Baxters Bus Services of Airdrie took this one Reliance in 1958, becoming their fleet no. 112. It is seen in the Depot yard at Airdrie. Baxters two tone blue and grey livery looked very smart on these vehicles. This style of body was a 'halfway house' between the curved one above and the straight one below having a straight roofline but a curved waistline.

196. Alexanders took 19 more Aberdonians in 1958 which brought their total up to 22. They too were seated for 41 and NL4 is seen on private hire duties in Aberdeen.

197. An odd body went to Garelochhead Coach Services on an AEC Regent V chassis in 1958. HSN485 is seen outside Helensburgh railway station in 1960. Note the 'clippie' with typical independent 'uniform' of dustcoat and her insert Setright register, common too with Scottish independents as well as the SMT group..

198. Glasgow Corporation took 50 Leyland PD2/24 chassis in 1958-9 and these were fitted with 61 seat Alexander bodies. L241 is seen outside the Royal Infirmary heading through the City on the 37 route from Springburn to Castlemilk.

199. Baxters turned to Leyland for their next purchase and 113 was a Tiger Cub with 41 seat body. It was disposed of to Alexanders (Midland) in 1963 after the Baxter takeover by SOL. It became MPD265 and lasted in the Midland fleet until 1974, and was latterly equipped for OMO as seen here at Stirling.

200. Perhaps the most unusual bodies turned out in 1958 were two for Aberdeen Corporation on reconditioned Daimler CVD6 chassis. They were the last half-cab single deckers ever bodied by Alexander and combined some traditional detail with features of the 1957 bus body. Both vehicles were new in 1947 with Walker bodies. 35 coach seats of the style being used in the Scottish Bus Group's dual purpose vehicles were fitted and the vehicles were used by ACT on their City tour. No. 9 shown here in normal service finished up with Greyhound of Arbroath whilst sister 11 went to McLennan of Spittalfield.

201. This nearside view of No. 9 is taken in the depot yard at King Street.

202. It is felt appropriate that more pictures of these fine vehicles should remind us of a long line of front engined single deckers bodied by Alexander. Fortunately we have a rear view of former Aberdeen 11 in Shore Terrace Dundee when running for McLennan of Spittalfield. The rear end is almost pure 1958 coach design.

203. Four Tiger Cubs appeared in 1959 with 41 seat bodies for Alexanders. PD139 is seen here tucked away in Perth garage in original condition and livery.

204. The milestone event of 1958 was the bodying by Alexanders of one of the early Leyland Atlanteans. After exhibition at the 1958 Commercial Motor Show it joined the Glasgow Corporation fleet as LA1. This shot shows it in later life in the revised livery and in the training fleet from whence it passed into preservation and can now be seen in Glasgow's Museum of Transport in the Kelvin Hall.

205. Edinburgh took 5 Leyland PD3s in 1959 of which 264 was one of the first four which were PD3/3s. 264 is seen here and this displays an experimental livery which it kept until 1962. It is seen operating the 19 circle route; note the very upright rear dome fitted to these vehicles.

206. Glasgow Corporation had yet to move into the 30' era and their 1959/60 deliveries were split between Daimler CVG6 and Leyland PD2/24 chassis, all 61 seaters. One of the Daimlers, D222 is seen climbing north up Renfield Street before passing the Transport Offices on the corner of Bath Street.

207. Leyland L249 was the first of the Leylands and is seen in Renfrew Street at the City terminus of route 28 to Milton. Note the pull-in ventilators on the top deck and sliding ones on the lower deck, a feature of Glasgow specifications at the time.

208. Western SMT's intake of 1959 included 7 Bristol MW6G chassis with their standard version of the Coach body which retained a curved waistrail. These were 41 seaters and Ardrossan's T1492 is seen here.

209. There were also 38 of these PD3/3s for Western SMT and 1528 is seen in Ayr bus station. The remaining PD3s that year had Northern Counties bodies.

210. In 1960 Western returned to the Guy Arab UF fitted with the standard 41 seat body. 1553 is seen on home territory in Whitesands, Dumfries.

211. Scottish Omnibuses 1959 deliveries of AEC Reliances featured this body style not seen on any other vehicles. They were designed for the 'two day' London service and had 38 seats. There were 20 of these which could also be available for tour and private hire work. B673 is seen on the 'Two day' service parked up at Gloucester Green, Oxford, whilst passengers and crew enjoy a lunch stop.

212. There were 20 of these AEC Reliances bodied for Alexanders in 1960 but one was diverted to Scottish Omnibuses as their B775 for exhibition at the 1959 Scottish show. It should have been AC153. AC154 is seen here in Aberdeen when allocated to Rosehearty garage, in original Bluebird livery. They were dual purpose 41 seaters.

213. Tiger Cubs also featured in Alexanders 1959-60 deliveries and there were 30 in the first batch. MPD157 is seen leaving Dundas St. bus station Glasgow, on the ex Lawsons Lenzie via Whitegates service in Midland days, painted in bus livery. It survived until 1974.

214. There was a further intake of 20 Leyland PD3/3 chassis by Alexanders in 1959 and these were bodied in the now standard 67 seat lowbridge style. MRB224 heads out of Glasgow on the local service to the housing schemes of Drumchapel after the 1961 split, and shows the short lived wide ivory band applied to some of these vehicles in 1962.

215. Edinburgh 999 was a PD3/2 which made up the fifth of Edinburgh's 30' long double deck intake for 1959. When new it had an all scarlet livery but it is seen here on the 19 Circle in the more restrained madder and white in September 1961.

216. Aberdeen Corporation were not a regular Alexander customer until after 1959 when they took delivery of 5 AEC Regent Vs with 66 seat bodies. 2745 is seen in Union Street.

217. Another new customer was Sheffield Corporation who sent 20 AEC Regent V chassis for bodying in 1960. 867 displays the cream and dark blue Sheffield livery. Note the additional emergency exit required on 30' long vehicles incorporated into the front offside lower deck window.

218. Alexanders first Atlanteans en force involved 17 chassis for the Northern group in 1960. One of those supplied to Sunderland District as their 302 is seen on home ground. Sunderland District's blue and cream livery wore well on the early Alexander bodies for rear engined double deckers; the Alexander design was more curvaceous than most others in those early days before wrap around screens.

219. One of Gateshead's intake, No. 80 is seen outside Gateshead railway station on the local route to Wrekenton. Gateshead and District was an all tram operation until abandonment in 1951.

220. Weardale Motor Services of Frosterley took this solitary Leyland PD3/1 in 1959 to which was fitted a body to Edinburgh specification apart from certain minor details. It is seen in Bishop Auckland before heading west into the hills to the market town of Stanhope.

221. Central SMT stuck to the PD2 even when larger vehicles were available and a further batch of 15 were bodied by Alexander in 1959, a further 10 going to Northern Counties. L620 is seen heading back towards Glasgow (Waterloo Street).

222. Often made to be content with second-hand vehicles from other SBG members Highland Omnibuses did get a few new ones! B22 was one of 6 AEC Reliances new in 1959 and it is seen in the 'far north' at Wick depot.

223. Dundee Corporation sent 10 Daimler CVD6 chassis for rebodying in 1960 and these were the last to feature the 5 bay design on a short chassis. The vehicles originally had Barnard bodies when new in 1949; they lasted in service until between 1972 and 1974. 119 is seen at the Corporation stances below the Caird Hall on Dock Street..

224. Barton took a further 6 AEC Reliances in 1959. They, like Western SMT, continued to take the curved waistrail design. 810 is seen far from home in Kilmarnock on a Football charter. Bartons had a Scottish connection due to their Corby-Glasgow service and maintained an office in Glasgow at 9A Burgher Place from the early 1960s.

225. Four Tiger Cubs for Alexanders started the 1960 orders. MPD171 is seen after 1961 when allocated to Oban depot, working an Oban local to the Corran housing scheme.

226. Starks L15 came on the end of the Alexander order and was identical in all respects apart from the external livery. It is seen in St. Andrews Square bus station before Starks became part of Scottish Omnibuses, but as with all vehicles employed on the joint Edinburgh to Dunbar service, it has Starks own version of SOL livery.

227. The 1960 Reliances for Alexanders had a revised frontal treatment with reshaped grille and half bumpers. NAC146 is seen in Northern coach livery in Inverness. At the time it was an Elgin based vehicle.

228. Smith of Barrhead took a solitary Leyland PD2/30 in 1959 - it was seated for 59. It is seen here in Paisley on Smith's local service to Todholm.

229. Garelochhead Coach Services received this AEC Regent V in 1959. It is seen on Helensburgh front heading for home.

230. Newcastle Corporation took its first Alexander bodied Atlantean in 1960 and 187 seen here was also numerically the first of 52 delivered over the next two years, 25 having Weymann bodies and the remainder Alexander. It is seen on route 22 which will take the bus over the Tyne into Gateshead and the village of Wrekenton.

231. Sheffield also took its first Alexander bodied Atlantean in 1960. 369 is seen here near the City centre.

232. Scottish Omnibuses 1960 deliveries included a batch of 33 AEC Reliances. Five of the first sanction were diverted to Highland Omnibuses as seen below, whilst the remaining 28 entered service in March 1960. A further 5 followed in June. B785 is seen in Germiston Street, Glasgow, just outside the entrances to Killermont Street bus station. Note the use of the 'Scottish Omnibuses' title and the SMT diamond.

233. The 5 for Highland were identical except for external livery. B26 is seen in Inverness.

234. The solitary Alexander bodied Atlantean for Belfast Corporation, 551. It is seen in Great Victoria Street.

235. Western SMT split its 1960 deliveries between Bristol MW6G models and Leyland Leopards. One of the Bristols is seen here 'plying for trade' outside the Ardrossan office - an evening mystery tour is on offer. There were 13 of the Bristols, which had 41 coach seats.

236. *The remainder were on Leyland Leopard L1 chassis - Western's first. These were 30 seaters with toilets for the London service. 1613 is seen on an Omnibus Society Scottish Branch excursion to the Tramway Museum at Crich. The driver is recorded as having told his passengers that this was the first time he had ever used first gear on one of these vehicles.*

237. *Alexanders also took more Tiger Cubs in 1960 and one of the batch, PD183, is seen leaving Wembley after an International match. Compared with today's vehicles, it will be a long journey home to Alloa!*

238. *The shape of things to come! The 6 Albion Nimbuses delivered to Alexznders plus a further 4 to Lawsons featured this style of body with rear end styling something akin to the then current Ford car range - Anglia and Classic. As we will see in a future volume, this design was to be developed for full size chassis in 1961 and then disappear due to a fire which destroyed all the moulds. It is hoped to continue the Alexander (Coachbuilders) story in a further volume in this series.*

Photographic Credits

I am again extremely grateful for all those who have helped illustrate this volume. It is sometimes difficult to attribute certain shots to a particular photographer and if we err, please accept our apologies.

Alan Cross. Page 14 upper left and bottom, and Plates 2, 8, 9 and 14,

Alistair Douglas. Page 3, page 14 centre, all on page 15 and Plates 48, 52, 70, 91, 95, 100, 110-2, 115, 117, 122-3, 126, 132, 148, 154, 166, 191-2, 202, 210, 213, 224, 233, and 235-6,

J. B. Fulton courtesy Alistair Douglas. Plate 4.

Robert Grieves. Contents page, Page 14 upper right and Plates 29, 36, 80, 164, and 195.

David Harvey Collection. Plates 16, 17, and 105.

Bob Kell. Plates 133 and 230,

Roy Marshall. Plates 15, 18-21, 23-28, 30, 31, 44, 47, 50, 57, 58, 60-1, 63, 71, 73, 75, 77, 81, 83, 89, 97-9, 102, 113-4, 119, 127, 129, 130, 136-7, 144-5, 149-53, 158-60, 163, 165, 170, 172-3, 176-86, 189, 194, 197-8, 204-5, 207-9, 215-21, 228-9, 231 and 234,

Harold Peers. Plate 59.

Ray Simpson. Plates 3, 5, 10, 11, 12, 33, 35, 37-8, 40-3, 45-6, 49, 53-6, 62-9, 72, 74, 76, 78-9, 82, 84-8, 90, 92-4, 96, 101, 104, 106, 107-9, 116, 118, 121, 124-5, 131, 135, 138-43, 146, 155-7, 161-2, 167-9, 174-5, 187, 190, 193, 196, 199-201, 203, 211-2, 214, 222-3, 225-7, 232, 237 and 238.

Jim Thomson. Page 3 and Plates 103, 120, 134, 147, 171, 188, and 206. Colour slides on front cover and LA1 on rear cover..

Peter Walton. Plate 51,

S. N. J. White collection. Plates 1, 7, 13, 22, 32, 34, and 39,

Brian Wright. Plate 6.

The colour drawings are by the author. Details of copies of these for framing and over 800 other subjects can be had by writing to the Publishers.

Bibliography & Further Reading.

Readers will find that some of the books listed are out of print but are available second-hand from good suppliers. In researching and writing this book primary sources have, in the main, been consulted and previously published works used to check facts etc. afterwards..

Booth, Gavin...Alexander Coachbuilders, Transport Publishing Company 1980

Brown, Stewart J....Alexanders Buses, Roadliner Transport Books 1984.

Grieves, Robert...Alexander Album, XS Publications 1978.

Condie, Allan T. Alexanders Buses Remembered Vol 1. 1945-61. Allan T. Condie Publications 1996.

Condie, Allan T. Alexanders Buses Remembered Vol 2. 1961-85. Allan T. Condie Publications 1997.

The appropriate Fleet Histories , The PSV Circle and Omnibus Society.

PSV circle body lists BB189 and BB190.

Leyland Motors General Manager's minutes courtesy British Commercial Vehicle Museum - abstracted by David Bailey.

Afterword.

Whilst this book will, as a single volume, interest those with particular memories of the 1943-60 era, the volumes published to date are intended to be used as a set as some information carries over from the first two books.

It was felt that in this series of books that complete fleet or body lists were inappropriate as that information is available elsewhere, however it is the intention of the Alexanders Study Group to provide this information in due course, at least for the pre 1961 period. This publication does not include a complete body listing - we refer readers to the PSV circle body lists BB189 and BB190 for this information.

The Alexanders Study Group.

Has been formed to bring together all those interested in Alexanders buses and the products of . Alexander Coachbuilders. Details can be obtained from the publisher by writing or phoning the address on the contents page 2, of this book. A quality bulletin, to the same standards of printing as our books, is published occasionally, the first issue is still available.

Future Plans.

The first two volumes have covered the Alexanders story from 1945-85, and this volume the products of the Coachbuilding concern from 1945-60. In the fullness of time we intend to cover the whole of the Scottish Group of bus companies, concentrating on the 1945-60 period initially, but a book on Alexanders before 1945 has not been ruled out. Further volumes are likely to complete the Alexander bodybuilding story, at least into the 1980s.